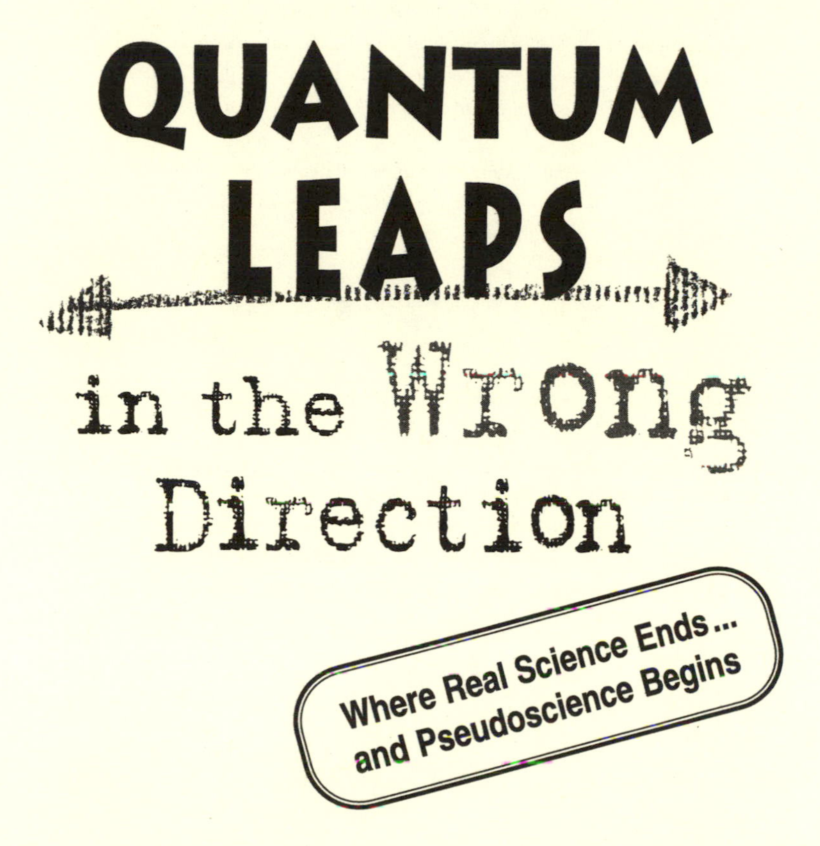

With Cartoons by Sidney Harris

JOSEPH HENRY PRESS
WASHINGTON, D.C.

Joseph Henry Press • 2101 Constitution Avenue, NW • Washington, DC 20418

The Joseph Henry Press, an imprint of the National Academy Press, was created with the goal of making books on science, technology, and health more widely available to professionals and the public. Joseph Henry was one of the founders of the National Academy of Sciences and a leader in early American science.

Library of Congress Cataloging-in-Publication Data

Wynn, Charles M.
Quantum leaps in the wrong direction : where real science ends and pseudoscience begins / Charles M. Wynn and Arthur W. Wiggins ; with cartoons by Sidney Harris.
p. cm.
Includes bibliographical references and index.
ISBN 0-309-07309-X (alk. paper)
1. Pseudoscience—Popular works. 2. Science—Methodology—Popular works. I. Title: Quantum leaps. II. Wiggins, Arthur W. III. Harris, Sidney. IV. Title.

Q172.5.P77 W96 2001
501—dc21 2001024426

Printed in the United States of America.

About the Authors

Charles M. Wynn, Sr., graduated from the Bronx High School of Science and the City College of New York and then attended the University of Michigan, where he received a Ph.D. in chemistry. After receiving his degree, he served as a Peace Corps volunteer at the Malayan Teachers College in Penang, Malaysia. He is currently Professor of Chemistry at Eastern Connecticut State University. He lives in Columbia, Connecticut, with his wife and three rabbits.

Arthur W. Wiggins graduated from the University of Notre Dame and then attended the University of Michigan, where he received an M.S. in physics. He is currently Professor of Physics and Physical Sciences Department Head at Oakland Community College in Farmington Hills, Michigan. He lives in Bloomfield Hills, Michigan, with his wife, two cats, and a dog.

Professors Wynn and Wiggins are the co-authors of *The Five Biggest Ideas in Science.*

Sidney Harris is "America's premier science cartoonist" (Isaac Asimov). He attended Brooklyn College and the Art Students League of New York (City). He has published more than 600 cartoons in *American Scientist* and was elected an honorary member of Sigma Xi. An exhibit of his cartoons and paintings has been touring museums around the country since 1985. His cartoons have appeared in numerous magazines, including *The New Yorker.* He is the author of *49 Dogs, 36 Cats, & a Platypus: Animal Cartoons* (1999), *Einstein Atomized: More Science Cartoons* (1996), and a number of other books; he also illustrated Wynn and Wiggins' last book, *The Five Biggest Ideas in Science.* Harris lives in New Haven, Connecticut, with his wife and is thinking about purchasing a parakeet.

Contents

Prologue

Planet Earth about to be recycled. Your only chance to survive—leave with us.

Marshall Herff Applewhite

In early April 1997, the world was stunned to learn that a group of 39 people had committed the largest mass suicide in U.S. history in their communal home in Rancho Santa Fe, California. Dressed in black pants, flowing black shirts, and new black Nike sneakers, their faces hidden by purple cloths, they had ingested a lethal dose of barbiturates mixed with applesauce, enhanced by a shot of vodka, and then helped along by the asphyxiating effect of a plastic bag over the head.

Why, the world asked, did a group of seemingly intelligent individuals, possessing marketable skills, and comfortably housed in an upscale neighborhood, decide to kill themselves? They did it because of their belief that by committing suicide in this manner, they would shed their bodies, or "earthly containers," and be whisked away by extraterrestrials to a spaceship and a higher level of existence. Unfortunately for them, their belief was pseudoscientific: it was erroneously regarded as scientific.

And how did they arrive at this misguided belief? They arrived at it in a manner characteristic of many pseudoscientists: they received it from a charismatic leader, a man named Marshall Herff Applewhite. The "classmates," as they called themselves, blindly and tragically accepted the teachings of someone whose deep-seated ideas about the universe were erroneous. Applewhite had convinced them of the existence of a gigantic alien spaceship, said to be following a comet that had been named Hale-Bopp (after the two astronomers who had first sighted it in July 1995). This spaceship was to take them home to the "literal Heavens."

Let's compare the claim of Hale and Bopp two years earlier, that a comet was heading our way, and the claim by Marshall Herff Applewhite, that a gigantic alien spaceship was heading our way.

Comets make exciting and dramatic viewing: a moving celestial object consisting of a head and a luminous tail that points away from the Sun. To test Hale and Bopp's claim that the comet existed, other scientists aimed their telescopes at the location in the sky provided by Hale and Bopp. They too observed this comet. Eventually, the comet came so close to our planet that it was possible for people to see it with the unaided eye.

The prospect of sighting a gigantic alien spaceship would also be exciting and dramatic. In fact, two members of the Heaven's Gate commune decided they'd like to see the spaceship for themselves. In January 1997, when the comet could not be seen readily with the unaided eye, they purchased a telescope capable of providing a clear image of the comet. With this telescope, they observed the comet, but were unsuccessful in their attempt to observe the supposed spaceship. They then returned the telescope to the shop where they'd purchased it.

Instead of deciding that their evidence did not support a belief in an alien spaceship, these people decided that they didn't need physical evidence. They discarded the telescope—but not their belief. *Clinging to this belief cost them their lives.*

To understand what's wrong with pseudoscience, we'll first examine what's right about real science, and then be in a position to compare science's approach to reality with that of pseudoscience. We'll learn that science's most basic value is that all ideas about reality are subject to both testing by experiment and challenge by critical rational thought. Scientifically literate thinkers accept ideas tentatively. They base their acceptance on evidence rather than on authority. People who are not scientifically literate are more likely to accept ideas absolutely. They are more vulnerable to deficient or bogus ideas as put forth by charismatic leaders or charlatans.

We'll examine in some detail the five most widely believed pseudoscientific ideas along with several dozen other ones, and see how they stand up to scientific scrutiny. In an epilogue, we'll suggest ways to become a better scientist—and avoid becoming a pseudoscientist. We'll also supply a glossary of interesting terms related to the study of pseudoscience.

Three groups of people will read this book. One is largely unfamiliar with the phenomena we discuss. We hope these people gain useful insights while exploring unfamiliar territory. Members of the second group are already acquainted with the phenomena and already in agreement with our conclusions. We hope they gain new insights into what for them is familiar territory.

The third group consists of people already acquainted with the phenomena and already in disagreement with our conclusions. Will members of this group change their views as a result of reading this book? We hope so, but we also realize such changes face significant obstacles. Once people acquire a belief, they tend to adhere to that belief, *even in the face of contradictory evidence.* Explanations developed to explain phenomena become fixed, *even when those explanations are shown to be irrational or based on wrong evidence.*

This unreasonable resistance to change is known as belief perseverance. A useful strategy for overcoming the tendency of people to continue to seek out and find confirmation of their beliefs is to help them focus on disconfirmations, potential flaws in the reasoning that led them to the original belief. By drawing people's attention to contrary reasons, and then encouraging them to spell out (ideally, write down) contradicting reasons, the tendency to neglect contradicting evidence can sometimes be overcome.

Making such evidence more conspicuous helps eliminate several natural human biases: favoring positive rather than negative evidence (favoring reasons "for" over reasons "against") and disregarding evidence inconsistent with or contradictory to the belief. To this end, we have developed and make extensive use of a comprehensive list of potential flaws in the reasoning process leading to beliefs about phenomena.

To help us keep our sense of perspective, Sidney Harris will provide humorous insights in the form of his inimitable cartoons.

Willimantic, Connecticut C.M.W.
Bloomfield Hills, Michigan A.W.W.
New Haven, Connecticut S.H.

RESEARCH INSTITUTE
UNANSWERED QUESTIONS
UNQUESTIONED ANSWERS
S. Harris

QUANTUM LEAPS in the Wrong Direction

SCHOOL OF OLOGY
ANTHROP 301
ARCHAE 126
BACTERI 109
BI 326
ENTOM 217
ETYM 221
GE 204
PALEONT 113
PHYSI 312
PSYCH 209
TOXIC 307
S. Harris

1 The Road to Reality: Scientific Method

Science is built up with facts, as a house is with stones. But a collection of facts is no more a science than a heap of stones is a house.

Jules Henri Poincaré

For thousands of years, people have sought to understand natural and artificial (humanly created) phenomena occurring in the universe. In the attempt to explain these phenomena, a variety of fields have evolved:

anthropology	creationism	history	palmistry
astrology	divination	homeopathy	phrenology
astronomy	dowsing	iridology	physics
biology	geography	magick	psychology
chemistry	geology	numerology	sociology

The fields can be divided into two distinct groups:

anthropology	astrology
astronomy	creationism
biology	divination
chemistry	dowsing
geography	homeopathy
geology	iridology
history	magick
physics	numerology
psychology	palmistry
sociology	phrenology

The left-hand column is a list of SCIENCES that systematically study phenomena and try to understand those phenomena in a general way. The right-hand column is a list of fields that also study phenomena and try to understand them in a general way. These fields, however, do not qualify as sciences.

To understand why members of the right-hand column are not true sciences, we'll first examine the activities that characterize truly scientific endeavors. Then, we'll contrast these with the activities of false (or pseudo) sciences and see how and why they differ.

Scientific Method

Science can seem mysterious, especially when presented in great detail. In essence, however, it is remarkably straightforward. Scientists simply try to gain a fundamental understanding of natural phenomena.

Everyone uses scientific reasoning to some degree. For example, if you hear a noise in the middle of the night, it may be important that you understand the cause of the noise. You might conjecture that the noise was caused by your dog Domino chasing your cat Puck. That scenario might seem harmless enough to you that you'd decide to stay in your nice warm bed. But, if you wanted to make sure, you would get out of bed, turn on some lights, and look for evidence such as an overturned lamp or guilty-looking animals.

Let's look at this example in a more systematic, yet extremely useful, way. Science begins with OBSERVATIONS: You have OBSERVED a noise in the middle of the night. If your general understanding, or HYPOTHESIS, about the cause of the noise is correct, you could PREDICT that it was caused by the dog chasing the cat. You perform an EXPERIMENT when you get up and look for evidence of such a chase.

If the result of the EXPERIMENT is not the one you've PREDICTED (both Domino and Puck are sleeping innocently), then your general understanding is clearly inadequate and must be reformulated or RECYCLED as a REVISED HYPOTHESIS.

If the result matches the PREDICTION, this supports (but does not prove) the validity of your HYPOTHESIS. After all, the lamp may have been knocked down by a burglar.

Each time a hypothesis withstands these tests, its credibility increases. Each time it does not, the hypothesis must be either revised or discarded. *Scientists must be open to either possibility.*

Here's another example. If you want to lose weight and think you understand your behavior well enough to choose an appropriate weight-loss technique, you test that understanding whenever you choose and then use a technique. If you do lose weight, the understanding of your behavior is intact. If you do not lose any weight, you have got to admit that your initial understanding was inadequate.

In this example, you have OBSERVED how you feel about your body, how you behave in the presence and absence of food, how often you exercise, and so forth. If your general understanding or HYPOTHESIS about your behavior is correct, you should be able to PREDICT which weight-loss technique (dieting by yourself, dieting and exercising by yourself, dieting as a member of a group that meets regularly, dieting using a plan monitored by your physician, etc.) most closely matches that behavior and will therefore most likely help you lose weight. You perform an EXPERIMENT when you actually attempt to lose weight using the chosen technique.

If the result of the EXPERIMENT is not the one you've PREDICTED (not only did you not lose weight, you gained weight!), then your general understanding or HYPOTHESIS about yourself is clearly inadequate and must be reformulated or RECYCLED as a REVISED HYPOTHESIS.

If the result is the PREDICTED one, this supports (but does not prove) the validity of your HYPOTHESIS. After all, you might also have lost weight using a different technique. It is important that scientists make every effort to be aware of any assumptions they make in formulating the hypothesis. If these are not valid, the experiment may not provide a valid test of the hypothesis. In the first example, the cat might have been chasing the dog. In the second, a woman who is not aware that she is pregnant might gain weight during the diet as a result of her pregnancy.

Another way scientists test hypotheses is by looking for preexisting (but as yet unknown to them) examples from reality that are consistent with their statement. For example, if you visit Disney World and observe that it rains briefly every afternoon during your week-long stay, you could evaluate the hypothesis that ***it rains briefly in the afternoon all year long*** not only by predicting a brief afternoon shower for tomorrow, but also by looking at local weather reports in the local newspaper for the past several months. If your search reveals a dry spell that lasted several days, the hypothesis will have to be revised accordingly.

Scientists thus have two ways to evaluate hypotheses: by seeking new instances predicted by the hypotheses, and by looking for preexisting examples consistent with the hypotheses. It is the obligation of professional scientists, as well as anyone who claims to use scientific reasoning, to continuously and relentlessly devise ways to employ these evaluation techniques. *If they do not, they risk clinging to false beliefs.*

Scientific Observations

Let's now take a closer look at how science observes and evaluates phenomena so that we can contrast this approach with that of pseudoscience.

Observations are the "facts" upon which hypotheses are based. Such facts become available when we perceive specific physical realities or events, such as noise levels measured on a sound meter or rain showers recorded by a rain gauge.

Scientific hypotheses or explanations must be based on observations of real phenomena. Most of the time, what we believe we sense is what actually occurs. If this were not so, we couldn't function effectively in the real world. Occasionally, however, our senses mislead us. For example, when we close our eyes after staring at the TV for a long time, the image of the TV screen is "still there"; our mind has played a trick on us by continuing to create an image from nerve signals received from the retina, even though the retina is no longer receiving light from the TV screen. *Events or phenomena may seem real but may not necessarily be real.*

Scientists have to keep in mind the limitations of personal experience when realities or events are sensed by human observers. For this reason, they need objective measurements rather than subjective ones. They seek repeated observations by independent observers. They seek observational evidence that is open to public scrutiny rather than guarded private information. They require corroboration of findings by other observers. Observations must be reproducible, so that any suitably trained observer will be able to sense and affirm their reality. *Scientists cannot allow authoritarian pronouncements to replace objective evidence.* Likewise, celebrity endorsements count only as personal opinions, not authoritative statements!

Furthermore, perceptions of reality can be influenced by prior beliefs or expectations. Perception—the act of knowing what our senses have discovered (light waves hitting our eyes, pressure waves vibrating structures inside our ears)—is the meaning or interpretation of these sensations as constructed by our minds. Since perceptions are learned, *there is a tendency for the mind to envision or construct what it expects to see.* For example, the minds of people who believe in and expect to see UFOs may construct images of UFOs from stray lights in the sky.

In essence, these people turn the statement, "I wouldn't have believed it if I hadn't seen it," into the statement, "I wouldn't have seen it if I hadn't believed it." Or, as written in the Talmud: "We do not see things as *they* are; we see things as *we* are."

Scientific Hypotheses

Sometimes more than one explanation is consistent with the observations. If no experimental evidence is available for making a choice among competing hypotheses, scientists select the simplest hypothesis as the one that is *most likely* to be correct. Scientists refer to this approach as Occam's razor, named after the

English philosopher William of Occam. They realize that the simplest explanation is not necessarily the correct one, but choose not to add complexity until they have experimental evidence that requires a more complex explanation.

Let's suppose you just attended a parent–teacher conference where you met your child's teacher for the first time. The conference was short and pleasant. That evening, while shopping in the supermarket, you see the teacher walking toward you. Instead of acknowledging you, the teacher just passes by without a word.

One way of explaining the teacher's behavior is to believe that ***he recognizes you but feels you were so rude to him at the recent meeting that he doesn't want to have anything to do with you.*** Another is to believe that ***he recognizes you but feels your comments were so immature or inadequate that he chooses to ignore your existence.*** Yet another is to believe that ***he is too elitist to speak to parents outside of school.***

How would a scientist explain the teacher's behavior? She would adopt the position that the most likely explanation is the least complicated one: ***he simply doesn't know you well enough after one meeting to remember your face.***

Occam's razor is summed up for medical students by the statement: When you hear hoofbeats, think horses, not zebras. In other words, a given set of symptoms should be diagnosed initially as the most likely disease that fits those symptoms, and not as some rarely encountered exotic disease. A patient exhibiting a low-grade fever, sniffles, and a cough is most likely suffering from a common cold and not smallpox! However, if other symptoms such as a speckled rash on the face and watery eyes appear a few days later, the patient may have a less common disease such as the measles.

To proceed from observations to a hypothesis, scientists use a form of logic called *inductive reasoning.* Inductive reasoning proceeds from specific truths to an uncertain general explanation. This type of reasoning does not lead automatically to a perfectly accurate hypothesis; it merely produces a hypothesis that has a reasonable likelihood of being correct. Therefore, scientists must be relentless in their evaluation of the hypothesis for they may need to revise it.

The more experimental support the hypothesis receives, the more probable it becomes. However, *no amount of experimental support can ever prove beyond a shadow of a doubt that the hypothesis is absolutely true.* On the other hand, if the experimental results don't agree with the prediction, the hypothesis must be regarded as false.

"THERE COULD BE ANY NUMBER OF CAUSES FOR THIS CONDITION. PERHAPS HE BROKE A MIRROR, OR HE WALKED UNDER A LADDER, OR SPILLED SOME SALT..."

Scientific Predictions

Scientific hypotheses are both explanatory and predictive. They help explain the general causes of what has been observed, while allowing forecasts of what should be observed.

To proceed from the hypothesis to a prediction, scientists use a form of logic called *deductive reasoning*. Deductive reasoning takes the hypothesis at face value and predicts what will happen (or might be discovered to have happened in the past) if the hypothesis is true. In a logical sense, the prediction is as valid as the hypothesis. It carries the truth (or falsity) of the hypothesis to the ultimate test, the experiment.

Scientific Experimentation

Although it is relatively easy to make predictions, it is often very difficult to conduct experiments to test them. Experimental variables must be carefully controlled and monitored. Potential bias on the part of the experimenter and subjects must be eliminated to every extent possible. Experimental conditions and results must be reported accurately so that other experimenters can compare results and resolve any discrepancies.

Scientific Recycling

From a logical standpoint, if an experiment is properly designed and the experimental results match the predictions, the hypothesis is supported (at least until it is tested again). If the experimental results do not match the prediction, the hypothesis must be revised, or even discarded. For this reason, *scientists cannot become too attached to their hypotheses.*

In reality, however, comparing experimental results and predictions can be difficult. It is not always easy to determine just how closely (within what margin of error) the results must match the prediction. For this reason, refinement of the prediction and further experimentation may be necessary to eliminate reasonable doubt.

Here is an overview of the reasoning process used to evaluate scientific ideas.

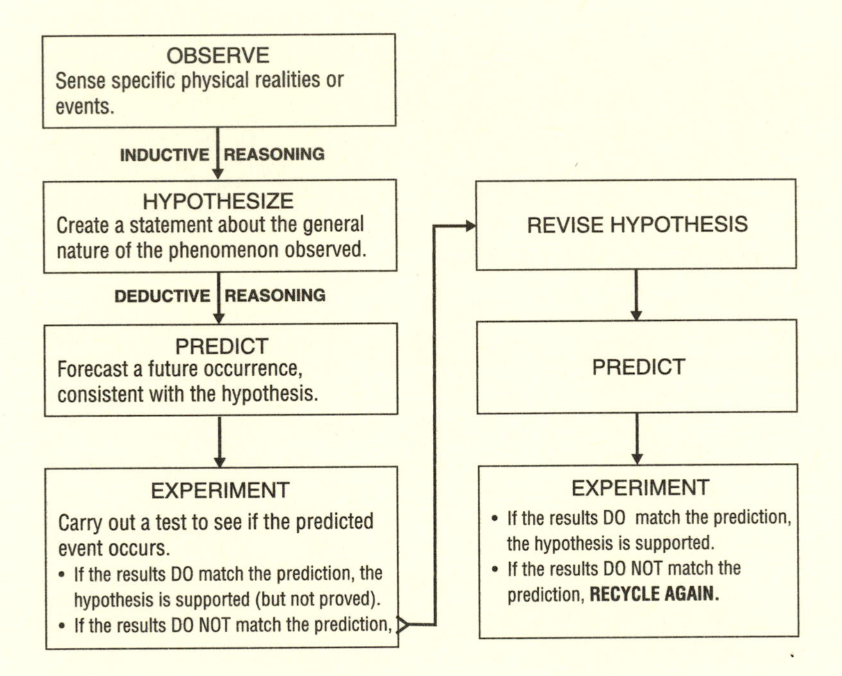

Hypotheses, Laws, Theories, and Models

Each time an experiment matches a prediction, the hypothesis gains credibility and dependability. After many successful tests, it may be called a *theory* (e.g., ***Einstein's theory of relativity***).

Theories frequently explain a *law*, which is a statement of some kind of regularity in nature (e.g., ***Newton's law of gravitation***). Theories might postulate the underlying cause(s) of the law's regularity. Another type of hypothesis is called a *model*, a representation or likeness of reality invented to account for observed phenomena (e.g., ***the plate tectonics model of Earth***).

"SINCE HE BECAME AN ARISTOTELIAN, HE HAS STOPPED EXPERIMENTING, AND HE NOW BELIEVES THAT ALL KNOWLEDGE COMES FROM THOUGHT ALONE."

2 Scientific Reasoning in Action

In science the important thing is to modify and change one's ideas as science advances.

Herbert Spencer

Evolution of Atomic Models

To see scientific reasoning in action, let's examine a classic example: scientists' quest to understand the unseen, basic building blocks of all matter. This example will show that scientific ideas develop not on the basis of authority, but through a rigorous refining process that compares reality to predictions. It will emphasize the need for scientists to continually reexamine their hypotheses in the light of new experimental evidence and to remain prepared to revise their hypotheses.

Democritus's Idea About Ultimate Structure

Belief in the idea that there is an ultimate underlying structure to all matter (i.e., that it cannot be subdivided indefinitely) was first stated in about 420 BCE by the Greek philosopher Democritus. Presumably, Democritus, while walking along a beach one day, observed that matter such as the sand on a beach appears continuous when viewed from a distance; up close the beach is seen to consist of individual grains. His intuition then led him to suggest that all matter must have a similar graininess. He thought, for example, that water in the ocean could be divided into smaller and smaller drops until one reached the level of "atoms" of water, which he envisioned as tiny, smooth, round balls.

Democritus's Idea About Ultimate Structure Versus Aristotle's Concept of Indefinite Subdivision

Democritus's conception was overshadowed for almost 2,000 years by that of another Greek philosopher, Aristotle (384–322 BCE), who thought that there is no ultimate underlying structure, that matter can be subdivided indefinitely. Aristotle's conception arose from a set of principles that to him were self-evident. If a survey had been taken at the time, people would likely have accepted Aristotle's idea over that of Democritus, in part because Aristotle's authority was preeminent.

The Scientific Revolution Provides a Way to Evaluate the Ideas of Democritus and Aristotle

During the seventeenth century, a fundamental change occurred in the way science operated: Experimental evidence was installed as the final arbiter of the validity of hypotheses. This revolutionary way of thinking assumed that no principles are to be taken as self-evident and that all scientific hypotheses must be subject to experimentation capable of determining the credibility of predictions based on them.

Dalton Agrees with Democritus

By 1803, the English schoolteacher John Dalton had observed that compounds, substances that consist of simpler substances known as elements, always contain these elements in the same proportions by mass—their composition by mass is constant. To explain this relationship, he used Democritus's concept of atoms and said that elements are composed of these extremely small, indestructible, indivisible particles. Dalton pictured these atoms as miniature billiard balls.

Dalton theorized that an atom of a given element has its own fixed mass. Dalton's theory enabled him to offer an explanation for the relationship among the masses of the elements in a compound. He reasoned that if a compound is characterized by a constant proportion by mass of its component elements, and each atom of a given element has the same mass, then the proportion by mass of the atoms—the compound's composition—must always be constant. (If the size of each combining unit was variable, the proportion by mass in compounds would also be variable, that is, not constant.)

It took about 2,000 years for scientists to accept an atomic model of matter. Dalton's model, however, is not the model conceptualized by today's scientists, for atoms are far more complex than billiard balls.

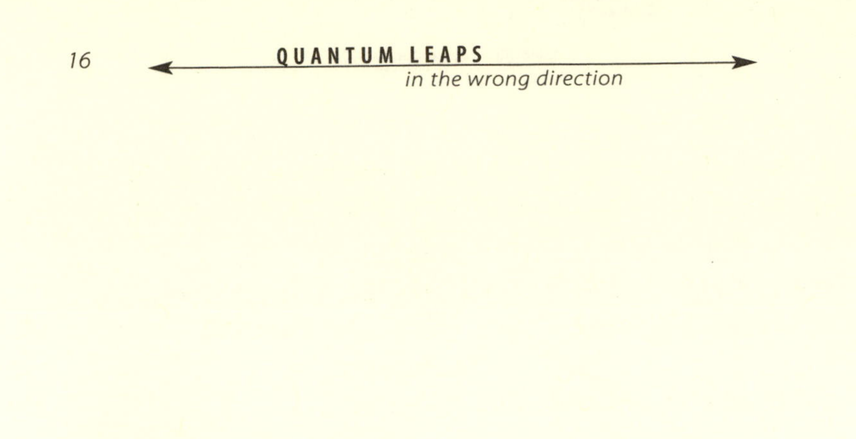

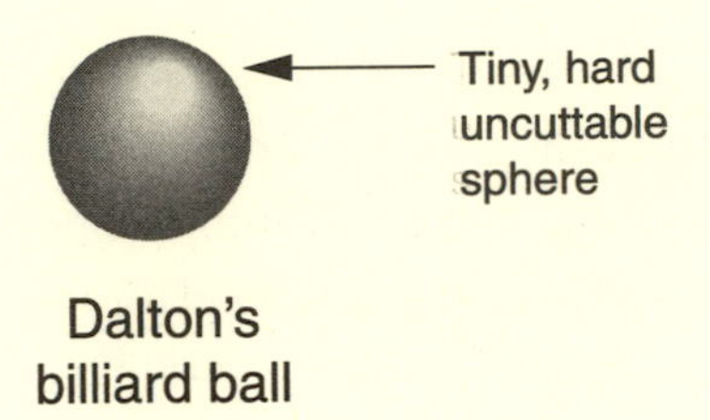

Thomson Adds Internal Structure to Dalton's Model

In 1897, the English physicist Sir J.J. Thomson, while working at the Cavendish Laboratory in Cambridge, England, gained evidence that all atoms contain negatively charged particles, which he called electrons. Since atoms were known to be electrically neutral, Thomson reasoned that there must be some positively charged material inside atoms to counterbalance the negative charges of electrons.

According to his hypothesis—the Thomson plum pudding model of the atom—an atom is spherical and consists of a thin cloud of positively charged material, with negatively charged particles embedded throughout, like raisins in plum pudding. Thomson's model was based on all known OBSERVATIONS about atoms. He employed INDUCTIVE REASONING when he used those observations as premises to support the supposed truth of his HYPOTHESIS.

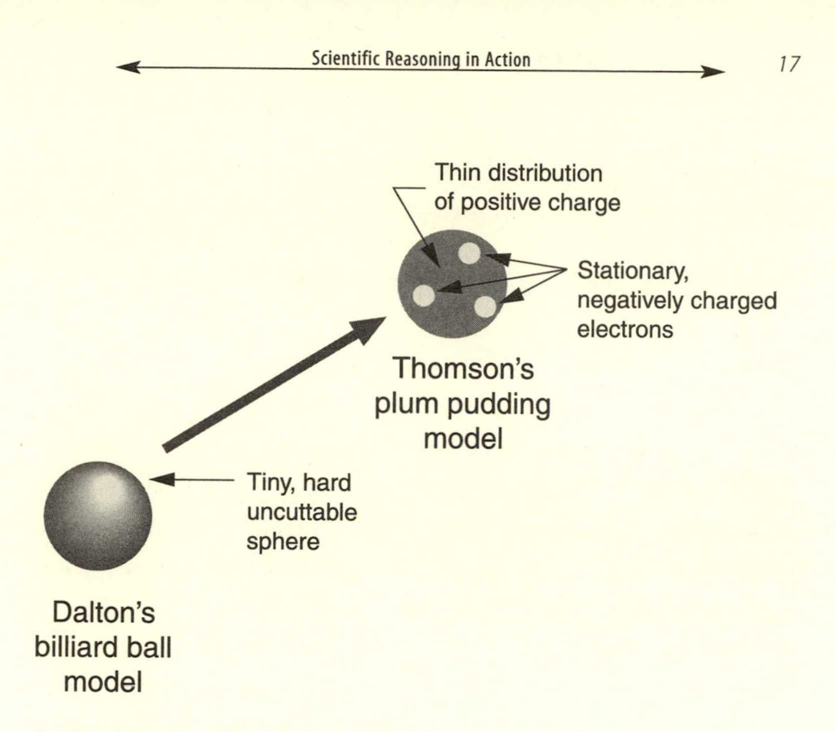

Rutherford Tests Thomson's Model

Thomson's successor at the Cavendish Laboratory, Lord Ernest Rutherford, a physicist from New Zealand, started from Thomson's model. Using DEDUCTIVE REASONING based on the premise of Thomson's hypothesis, Rutherford made a PREDICTION in 1910 about as yet unobserved phenomena. He reasoned that, if atoms consist of an insubstantial but positively charged "pudding" sprinkled with electrons, then these atoms would present little resistance to the passage of subatomic, positively charged particles (alpha particles, which are given off by naturally radioactive materials) projected directly at a thin foil made of gold (gold atoms).

Rutherford predicted that most of the particles would pass through unimpeded, but a small number would be slightly scattered as a result of repulsion by the wispy, positively charged material. The results of his EXPERIMENT *did not* agree with the predictions. Specifically, many more particles were scattered through larger angles than predicted.

Rutherford's Model Replaces Thomson's Model

Rutherford reasoned that the positive charge, instead of being spread throughout a sphere of atomic dimensions, was concentrated in a much smaller, extremely dense, centrally located region, which he called the atomic nucleus. Alpha particles that came close to this nucleus were deflected greatly by it and thereby scattered through large angles. He included this feature in his RECYCLED version of Thomson's plum pudding model.

This new model added the positively charged nucleus but retained the atom's spherical shape as well as the presence of the negatively charged particles. Because the electrical force seemed to hold the negatively charged electrons and the positively charged nucleus together in a manner reminiscent of the gravitational force that keeps planets orbiting the Sun, Rutherford

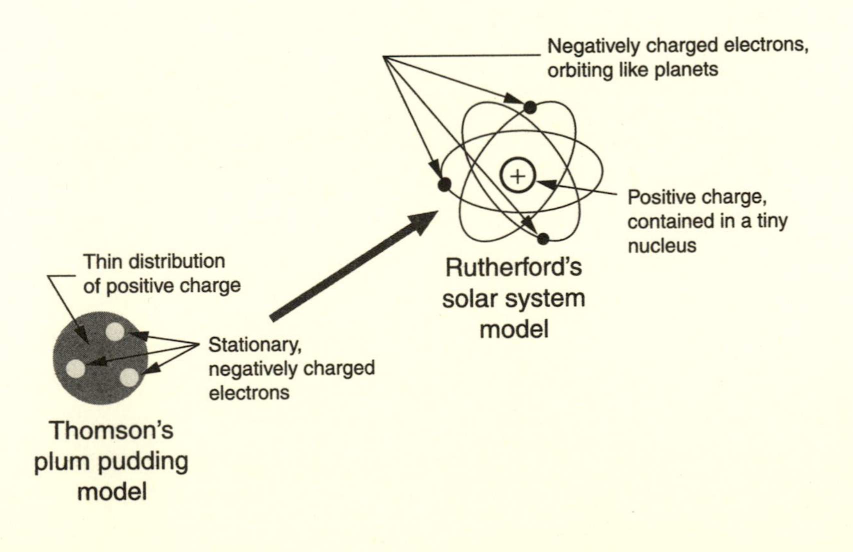

decided to depict the electrons revolving around the nucleus. Thus was born Rutherford's solar system model of the atom.

Other Models Replace Rutherford's Model

Although Rutherford's results supported his model, they did not (could not) *prove* that the model was correct. Certain features of his model, notably those having to do with the nature of electrons within the atom, were found inadequate to explain subsequent experimental results. Judicious recycling of Rutherford's model and its successors has led to today's quantum mechanical model of the atom.

Will the quantum mechanical model be the final one? Because of the nature of the scientific method, it is not (and cannot be) known whether any version will endure or will require additional recycling. *Even if a scientist happens to discover a hypothesis that is absolutely true, there is no way of knowing that he or she has done so.*

Here's an outline of the evolution of atomic models.

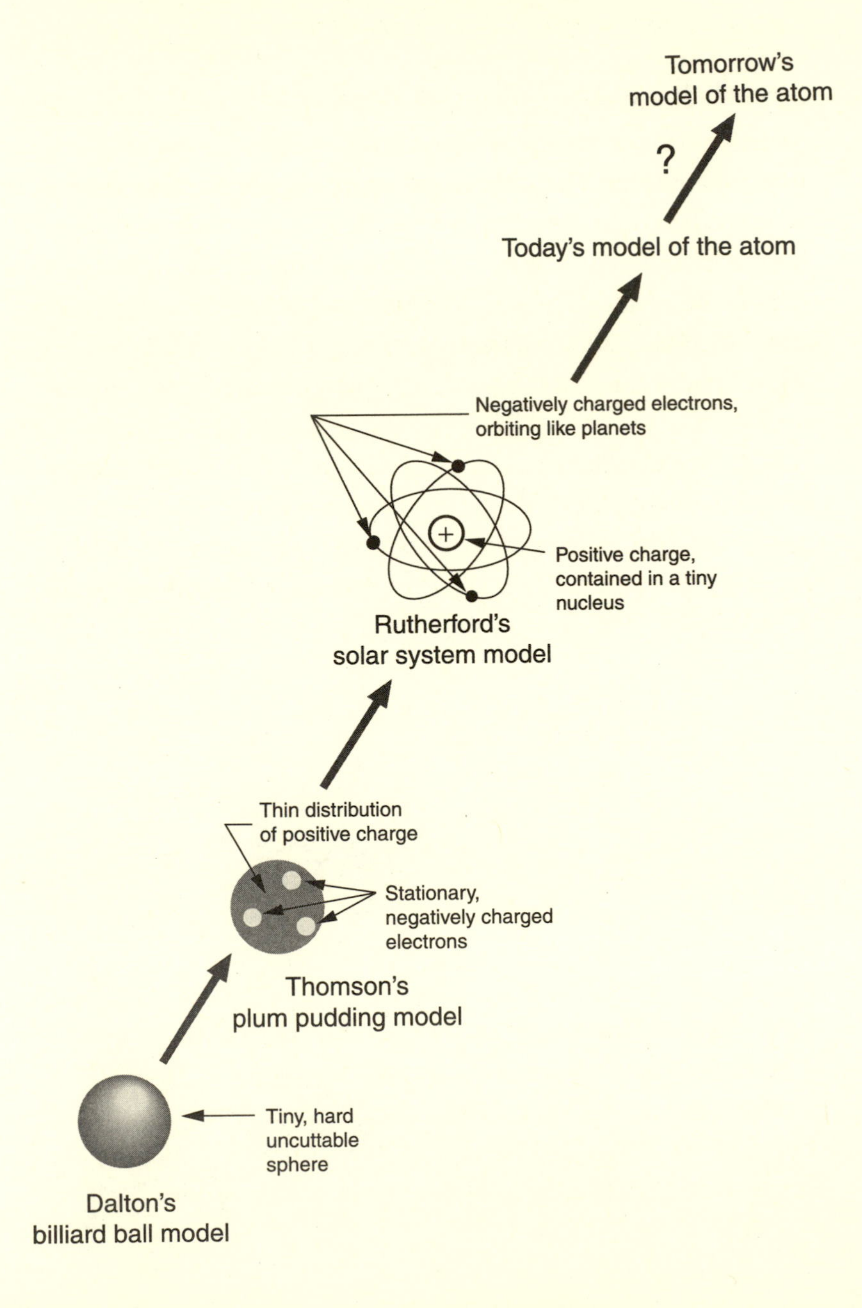

Nobody's Perfect

The rigorous process by which scientists like Dalton, Thomson, and Rutherford seek to understand the universe demands that all assertions about natural phenomena be subjected continuously to public scrutiny. The authority of science lies in its methods, not in individual scientists, who are fallible and can sometimes make honest mistakes.

This was true, for example, in the case of the French scientist Rene Blondlot. Blondlot became aware of the work of Wilhelm Roentgen, a German physicist. Roentgen was not looking for X rays when, in 1895, he covered an evacuated glass vessel with black paper to prevent the emergence of ordinary light from the vessel. After applying a voltage to the vessel, he noticed that a black line had developed on some photographic paper lying nearby. Roentgen believed this line had been caused by a new form of radiation. Subsequent careful testing by Roentgen supported this belief. He called these hitherto unknown rays X rays because, in mathematics, *X* usually connotes the unknown.

When reports of Roentgen's discovery reached Blondlot, he began to do experiments with these X rays. In one attempt to generate them, he chose a very hot platinum filament as the source of rays. The filament was enclosed in a sealed iron tube. A thin slit in a piece of aluminum allowed radiation to escape into the laboratory where its properties could be tested. Blondlot began to notice effects that were unlike those expected of X rays; for example, the luminosity of a nearby gas flame seemed to increase, and a screen painted with cadmium sulfide seemed to brighten.

Blondlot named the rays N rays, the *N* standing for Nancy, the home of his university. He sought other substances that would serve as N-ray sources. Iron and most metals seemed to emit

N rays naturally, but wood did not. Before the end of 1903, Blondlot had published 10 papers on the subject.

Just as Blondlot had replicated the X-ray experiments of Roentgen, other scientists tried to replicate the N-ray experiments of Blondlot. Although some scientists, such as Becquerel and Charpentier, claimed their experiments were successful, *many others were unable to replicate Blondlot's results.*

In 1904, an American physicist, Robert Wood, was sent to Blondlot's laboratory to investigate. Wood watched closely as Blondlot demonstrated a number of his experiments. In one, he used lenses of aluminum to focus the N rays and prisms of aluminum to disperse the rays onto a screen. He introduced a device designed to catch variations in the intensity of the N rays projected by the prism. When Blondlot used the device, he detected light and dark bands on the screen. When Blondlot allowed Wood to make the observation himself, *Wood could see no variation in the brightness of the mark.*

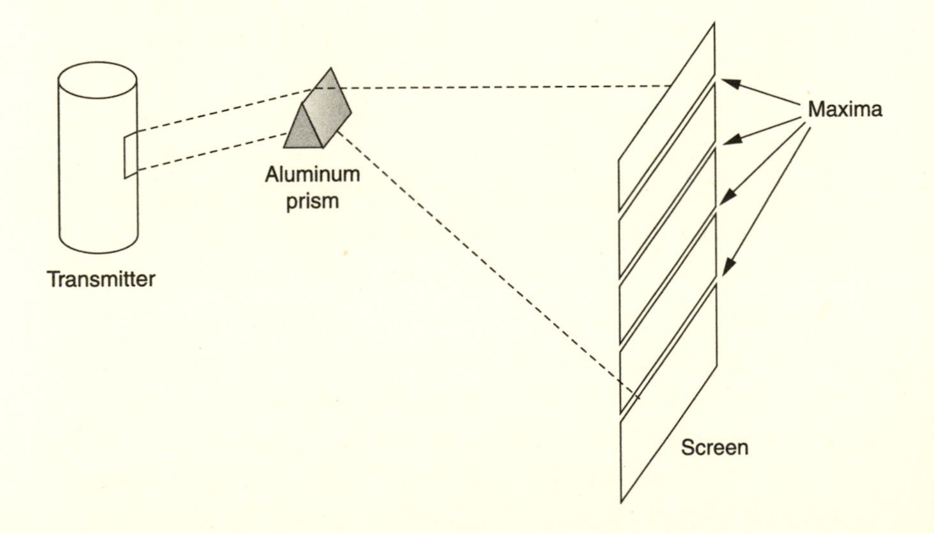

Wood then surreptitiously removed the aluminum prism that was alleged to disperse the rays. Blondlot continued to detect light and dark bands on the screen. In another experiment, Blondlot held a flat iron file just above his eyes. Blondlot said the N rays emitted by the file enhanced his vision and enabled him to see the hands of a faintly illuminated clock on the far side of the laboratory. In the darkness, Wood substituted a wooden ruler for Blondlot's file. Blondlot still saw the hands of the clock quite clearly, even though wood was not supposed to emit any N rays. Blondlot's N rays were extinguished after Wood filed a report of his visit in the British scientific journal *Nature.*

How could so many eminent scientists have been wrong? They had fallen victim to *perceptual construction.* In this phenomenon, people do things like connecting faint but distinct markings "in their minds" until the array seems to be a continuous line. "The face on Mars" is the result of perceptual contruction. The "face" was perceived from images received during the Viking mission to Mars in 1976. What the Viking orbiter had sent back was an image of a rock outcropping looking like a gigantic humanoid head staring into space from the surface of the planet.

The face on Mars is an example of *pareidolia*, a type of illusion or misperception involving a vague stimulus being perceived as something or someone. Other examples include the "Old Man of the Mountain" rock formation in the White Mountains of New Hampshire, which looks like a face when viewed from the side at a distance; the face of the "Man in the Moon" seen at the full Moon; and the image of the face of Jesus Christ with a crown of thorns seen in skillet burns on a tortilla cooked by a New Mexico housewife in 1978.

The N-ray problem was created because all the tests were based on subjective judgments. Instead of using instruments to gather objective data, people's observations of relative brightness determined the results. Such subjective judgments can be affected by belief or expectancy. Scientists require that experimental results be not only reproducible, but also independently verifiable before they are accepted as facts.

True but Strange

Democritus's idea that there is an ultimate underlying structure to all matter must have seemed strange to Aristotle and his followers. After the criteria for judging scientific ideas changed, the idea of atoms finally became acceptable because experimental evidence supported it. Likewise, all other ideas about the realities of the world must be evaluated and accepted or rejected on the basis of evidence, not on the basis of their seeming extraordinary or wonderful to a particular individual or group.

Some strange-sounding scientific ideas have withstood these tests. Here's a strange idea from a branch of physics known as quantum mechanics: In the realm of subatomic particles (the quantum realm), individual subatomic particles don't acquire some of their characteristics (e.g., position and velocity) until they're observed. In other words, subatomic particles do not seem to exist in a definite form until observers measure them! Bizarre as this "quantum weirdness" may sound, it has been confirmed repeatedly in rigorous tests.

Some people have misinterpreted this finding. They reason that since normal objects are ultimately made of subatomic particles, ordinary things also must be observed in order to exist. This conclusion is a quantum leap in the wrong direction because the properties of the whole are not the same as the properties of its parts.

Quantum theory is concerned only with what happens to *individual* subatomic particles. Quantum effects at the level of individual subatomic events are averaged out on the macroscopic scale. The Moon continues to orbit the Earth even when no one observes it. Science continues to have a claim on objective reality.

An additional misinterpretation of the finding that particles don't acquire some of their characteristics until they're observed by someone is that "ultimate reality is in the mind of the observer" or "thoughts can make anything happen." Neither of these ideas can be derived from quantum theory. This theory says *nothing* about the role of human consciousness or mental processes in the physical world.

Another true but strange idea comes from Einstein's theory of relativity: The elapsed time observed between two events is not absolute; it depends on the frame of reference of each observer. For example, if the two events are successive ticks of a clock aboard a spaceship traveling past Earth at great speed, an observer on board the spaceship would observe the ticks happening at the same location. From the perspective of an Earth-based observer, successive ticks of the clock aboard the spaceship would be occurring at different locations because of the motion of the clock-containing spaceship relative to the Earth-based observer. As a result, the Earth-based observer would record a longer time between ticks. This "time dilation" effect predicts that clocks moving relative to an Earth-based observer will seem to run more slowly than clocks fixed to the Earth.

Peculiar as this idea may seem, it has been confirmed experimentally. If two clocks are set at exactly the same time, and one remains on Earth while the other is taken for a ride aboard a jet plane, upon return to Earth, the jet plane-based clock shows less elapsed time than the Earth-based one. The effect becomes

significant (measurable) only at extremely high speeds. It is unnoticeable and of no significance to us in our daily lives.

Don't be misled by exaggerated anti-aging claims based upon this finding!

Science Lives in Subdivisions

There is such a variety of phenomena in the universe that scientists usually specialize in one or, at most, a few areas of study. In the broadest categorization of the sciences, there are two major subdivisions: natural sciences, which observe nonliving as well as living parts of the universe, and human sciences (behavioral and social sciences), which focus on humans as rational/emotional beings and the organizations or systems (political, economic, religious, etc.) created and shaped by them. The left-hand column below lists natural sciences; the right-hand one, human sciences.

astronomy	anthropology
biology	geography
chemistry	history
geology	psychology
physics	sociology

Other sciences would have to be included to complete each list, for example, natural sciences such as ecology and human sciences such as economics. Interdisciplinary sciences such as biochemistry and social psychology would also have to be included.

Similar but Not the Same

Although the natural and human sciences both seek general explanations of phenomena, there are important differences in the ways they pursue these explanations.

OBSERVATIONS Natural sciences observe physical and biological entities that are reasonably identical (atoms, bacteria, fruit flies) and relatively large in number compared to the personal and social realities observed by human sciences. There are trillions of trillions of carbon atoms in the universe, and they all exhibit the same chemical behavior; there are about 6 billion humans on Earth, and each is unique. A chemist can say, "If you've seen one carbon atom, you've seen them all." A psychologist would never say that about humans!

HYPOTHESES Since the entities observed by the natural sciences are reasonably identical and available in relatively large numbers, they are easier to isolate and present far fewer variables. Because of this, hypotheses in the natural sciences can often be reduced to a single acceptable hypothesis. On the other hand, multiple acceptable hypotheses are frequently the case in the human sciences (e.g., psychoanalytic theory vs. behavioral theory vs. cognitive theory in psychology). Natural science hypotheses are usually more exact (can be expressed using relatively simple equations), have a smaller range of possible error, and are more easily freed of bias or prejudicial assessment by the observer.

PREDICTIONS Because hypotheses in the natural sciences have a smaller range of possible error, predictions based on them have a smaller range of possible error.

EXPERIMENTATION In the natural sciences, it's easier to set up and control variables, the experiments are easily freed of bias and seldom involve direct ethical concerns (exceptions include nuclear weapon development and genetic engineering), and the behavior of entities generally is not influenced by the experiment itself. Observe carbon atoms as closely as you like, and they'll pay no attention to you. Stare at humans, and watch out—they may stare back!

Human behavior is especially difficult to study because it can be influenced by a wide variety of factors. *It can even be influenced by knowledge of the hypothesis being evaluated.* For example, if investors who learn of a rising bond interest rate are also aware of the hypothesis that ***rising bond interest depresses the stock market,*** they may decide to sell stocks they might have kept if they had been ignorant of the hypothesis.

RECYCLING Experimental results and predictions are more definitive in the natural sciences and therefore easier to compare than those in the human sciences. One crucial experiment in the natural sciences can substantially alter a hypothesis, as when Rutherford's test shot a big hole in Thomson's plum pudding model of the atom. Crucial experiments in the human sciences are rare.

SCIENCE
PSEUDO-
SCIENCE
S. Harris

3 The Road to Reality Versus the Road to Illusion

Extraordinary claims require extraordinary evidence.
Carl Sagan

Pseudoscience Sells

Theories that *claim* to be scientific must be held to the rigorous standards of science. To ensure that all theories meet these standards, it is essential that people be sufficiently scientifically literate. Unfortunately, the battle against pseudoscience is an uphill one. The public reads more about pseudoscience and the occult than about real science. Books about pseudosciences such as astrology sell millions of copies. In addition, the public is bombarded by pseudoscience in the form of TV dramas such as *The X-Files* and movies featuring gigantic alien insect invaders. These "special effects" can now be produced so convincingly that it becomes difficult to know where reality ends and illusion begins.

As a result, the number of people who are able to distinguish between science and pseudoscience is diminishing. More people believe in extrasensory perception (ESP) than in evolution. There are more astrologers than astronomers. A sign seen recently in a bookstore read: NEW AGE SECTION MOVED TO SCIENCE SECTION.

Increased belief in pseudoscience is a global trend. It responds to the search for personal powers we long for but can't seem to find. It promises relief from diseases. It even promises that death is not the end. It offers easy and immediate answers, as well as satisfying a craving for certainty. It serves powerful emotional needs and satisfies spiritual hungers. It promises to give people things that just don't exist. *The twenty-first century age of science is in danger of becoming the age of pseudoscience.*

Pseudoscience: Harmless Diversion or Harmful Fantasy?

Humankind's drive to gain insight into reality stems from two major motivations: our innate curiosity about the world and our desire to influence the human condition by controlling that world. *When fantasy replaces reality (when pseudoscience replaces science), our ability to know and influence the real world is diminished.*

While purveyors and consumers of pseudoscience believe they have much to gain from their pseudoscientific beliefs, in fact, they have much more to lose. In addition to the money they may invest in ill-conceived schemes or in support of con artists, they also invest time that could be more profitably spent expanding their knowledge of reality. Pseudoscientific medical beliefs can even harm them physically if they seek help from faith healers, psychic surgeons, and other medical quacks for potentially life-threatening problems. Even if they later turn to scientifically based medical treatment, it may be too late.

Attempts by religious fundamentalists to require public schools to present religious explanations of natural phenomena alongside or in place of scientific ones are especially dangerous. If the attempts are successful, students will be indoctrinated with pseudo-scientific beliefs and will leave school with warped and restricted views of reality. *And they in turn will teach the next generation!*

"CAN'T YOU GUYS READ?"

Natural Versus Supernatural

Scientists attempt to explain natural phenomena, as well as phenomena created and shaped by humans. They also attempt to explain supposedly "supernatural" phenomena that seem to violate the natural order of things but in reality have naturalistic explanations. *A phenomenon that has not yet been explained is not necessarily supernatural.*

For example, to the ancient Greeks, a hailstorm was one of the ways in which the god Zeus showed his anger. To modern meteorologists, a hailstorm occurs when an upward air current brings droplets of water into high, cold atmospheric layers where the droplets freeze. This can happen again and again, and the more often it happens, the bigger the hailstones are likely to be.

The scientific explanation for a phenomenon may already be available in terms of current theories, or it may require recycling of a currently held theory, as when Rutherford explained the surprising (to him) scattering of alpha particles by theorizing the existence of atomic nuclei. Similarly, chemistry's periodic law, which explains trends in the behavior of different kinds of atoms, was originally expressed in terms of the masses of different kinds of atoms, but is now given in terms of the number of positive subatomic particles (protons) in these atoms.

Astronomy's model of the solar system, which was once geocentric (Earth-centered), is now heliocentric (Sun-centered). Geology's early 1900s model of the Earth had difficulty explaining the apparent "drifting of the continents" until a mechanism for continental drift was provided by underlying sidewise forces generated by currents in the mantle layer that lies beneath the continents. The genetic mechanism that helped explain Darwin's theory of evolution did not become available until the structure of deoxyribonucleic acid (DNA) was determined.

Science and Magic

Scientists endeavor to explain all phenomena in naturalistic ways. This quest led the English physicist Sir Isaac Newton, for example, to formulate his ***law of universal gravitation***, which states that all objects in the universe exert a gravitational force of attraction upon each other. This law describes an invisible attractive force between apples and planet Earth, and predicts that baseballs hit into the outfield will eventually descend to Earth.

Before the relevant explanation or law has been discovered, such phenomena can seem to have supernatural or "magical" qualities. Thus science and magic are not strangers to each other. For example, there is an interesting rock called lodestone. By means of an "invisible" force, it has the power to attract iron from a distance. This invisible force was considered mysterious until scientists understood the phenomenon of magnetism and the natural laws that describe it (just as Newton had described the fundamental laws of gravity). Lodestones are simply naturally occurring magnets composed of the magnetic mineral magnetite.

A person who is unaware of the phenomenon of magnetism could be fooled by a magician who presents lodestone as a "magic rock," perhaps as a formerly ordinary rock made magical by saying the word *abracadabra*. When she pretends that the phenomenon is occurring because of her "magical" influence, the presentation becomes a magic "trick" or illusion.

Most people enjoy the seeming suspension of reality effected when magicians exhibit natural phenomena such as lodestone magnetism or resort to deliberate deception to give the illusion that natural laws are being defied. There is always an explanation for these tricks—but don't expect the magician to furnish it.

Rising to the Occasion

Here's an interesting natural phenomenon that you can present as magic if you wish. Pour fresh ginger ale or another light-colored soda into a tall glass. Drop a few purple raisins into the soda. Explain that these "magic" raisins usually obey your commands, but that some are more obedient than others.

Bubbles will begin to collect on the raisins. In a few seconds, raisins will start to rise. As soon as you see one begin to move, *command* it to rise. Then, when it reaches the surface, tell it to fall (and it will!). Of course, your commands have nothing to do with the rise and fall of the raisins. In fact, if you told them not to rise, they'd disobey.

The scientific explanation for this phenomenon is that soda contains carbon dioxide gas. In the absence of raisins, the gas simply collects as bubbles that rise to the surface. They rise because the buoyant force of the soda water is greater than the weight of the bubbles, which form on the many points of attachment on a raisin's rough surface. As they collect, the raisin becomes increasingly buoyant until it finally rises to the surface of the soda. As a bubble on the raisin rises, it expands as the pressure on it lessens. When it reaches the surface, it expands even more, stretching the film of liquid surrounding it until the film is stretched too thin to hold the gas inside, and the bubble breaks, releasing the gas to the air. With the loss of support from the gas bubbles, the raisin sinks to the bottom, where it remains until a new group of bubbles collects on its surface.

Magic versus Magick

Most professional magicians prefer to be called illusionists, to emphasize that they are only performing "tricks." The magic they perform involves deliberate, but admittedly deceptive, means to make it appear as if they have supernatural or paranormal powers. Such magic should not be confused with "magick," a pseudoscience that purports to willfully contravene the laws of natural science. Sai Baba in India performs magick when he pretends to materialize copious quantities of ashes from his hands. Uri Geller of Israel does likewise when he seems to bend spoons using only the power of his mind.

When we (the authors) taught together, one of us would explain to our class that science knows of just four naturally occurring physical forces: gravitational force, electromagnetic force, and two forces that act only within the nuclei of atoms, the strong nuclear force and the weak nuclear force. The other person would then announce that he had just discovered a fifth force, which he was prepared to demonstrate. He placed the palm of one hand on top of a book, wrapped the other hand underneath the wrist of that hand, and began to lift the flattened palm. Amazingly(!), the book also rose, as if attracted to the palm by this fifth force.

Most of the students, however, saw right through the trick: The unseen index finger on the hand holding the wrist is moved to a position underneath the book. Those that didn't get the trick at first knew enough about our hijinks to realize they were being tricked.

Pseudoscientific Observations

Let's now compare science's use of observations, hypotheses, predictions, experimentation, and recycling with that of pseudoscience.

Observations are the facts upon which hypotheses are based. Observational problems arise when bias on the part of an observer produces reports that do not correspond with reality. Wishful thinking causes people to imagine events happening that, in reality, do not happen, especially when those events correspond to strongly held beliefs. For this reason, people who rely primarily on personal anecdotes as evidence are in danger of deceiving themselves. They may, for example, tend to notice only positive events corresponding to a belief, while ignoring negative ones (as when a dowser locates a source of water only once out of many attempts).

Dishonesty in making and reporting observations is another potential problem. Honest reporting is a basic tenet of science. Fraudulent observations are relatively rare in science. When discovered, they are usually dealt with in a timely fashion. On the other hand, observers of phenomena not within the range of normal experience, so-called paranormal phenomena, have much more frequently been exposed as frauds or charlatans acting in their own interest.

Pseudoscientific Hypotheses

Occam's razor does not cut it with pseudoscientists. Rather than adopt the simplest explanation as a matter of principle, they embrace explanations that are so broad, vague, or changeable that they are rendered immune to scientific study.

Pseudoscientific hypotheses can be especially appealing if they respond to emotional needs such as a desire for easy and immediate answers, a craving for certainty, spiritual hungers, health concerns, and yearnings for an afterlife. Such explanations are often based on belief systems that demand faith in powers or forces for which there is no evidence, and, in the process, require believers to abandon well-established scientific hypotheses.

Another problem with pseudoscientific hypotheses is that they may be formulated in such a way that there is no conceivable way to test them experimentally. For example, someone may argue that the cause of their occasional strange behavior is a rabbit that is invisible to other people, accompanies them everywhere, and persuades them to behave in odd ways. The invisibility of the alleged animal renders it undetectable and thus immune to objective evaluation.

Such explanations are said to be nonfalsifiable. Their falsity cannot be determined by any conceivable test. For an explanation to be scientific, it must be falsifiable: Conditions must exist under which we would be willing to set aside the explanation. For example, ***Newton's law of gravitation***, which predicts that apples will fall from apple trees, would be falsified if an apple moved upward from an apple tree. *If no such conditions can be imagined, the explanation is not a scientific one.*

Pseudoscientific Predictions

If a hypothesis is true, *then* predictions derived from it should hold true. Thus it should be possible to use deductive logic to derive predictions from pseudoscientific hypotheses. As such, these predictions should lead to legitimate tests of the hypothesis. Unfortunately, pseudoscientific hypotheses are usually so general or vague that predictions deduced from them allow too wide a margin of error for adequate evaluation.

Pseudoscientific Experimentation

Pseudoscientific experiments fall prey to the same difficulties (bias, wishful thinking, dishonesty, etc.) involved in the original pseudoscientific observations and the creation of the pseudoscientific hypothesis itself. Since the predictions of pseudoscientists are predicated upon prior beliefs to which the pseudoscientists are committed, it is not surprising when they seem to find what they believe they will find. It is also not surprising when fraudulent or self-serving observations are followed by fraudulent or self-serving experimentation.

Pseudoscientific Recycling

Even when pseudoscientific experimental results do not match predictions, adherents may still cling to their original belief because of its powerful attraction. *The assertion of dogma closes the mind.*

Adherents may argue that a belief has been held by so many people for such a long time that it must be valid. They may also argue that these believers are sincere in their belief. *Popularity and sincerity, however, are not evidence of truth in any scientific sense.*

In addition, their ideas may be embellished with a so-called conspiracy theory that some agency is withholding information that supports their belief. For example, they may contend that the government is unwilling to release its collection of alien-life-form cadavers, thus rendering the alleged phenomenon impervious to evaluation.

Pseudoscience: A Summary

To make the distinction between science and pseudoscience clear, let's summarize the problems or flaws typically associated with pseudoscientific thinking.

OBSERVATIONS Objective sensing (seeing, hearing, etc.) of specific events or entities.

- Observers are not properly trained or equipped.
- Observers exaggerate, mistake, or imagine phenomena.
- Positive instances are emphasized; negative ones are ignored.
- Unsupported personal anecdotes are relied upon as primary evidence.
- Measurements are subjective rather than objective.
- Observations are not reproducible.
- Con artists make fraudulent claims.

HYPOTHESIS A generalization, as simple and direct as possible, related to these observations and/or the apparent cause(s) of these occurrences, which is expressed in well-defined words or mathematical relationships and which is consistent with previous experimentally supported hypotheses.

- Not expressed clearly enough for definitive predictions and experiments to be carried out
- More complex than observations warrant
- Created by people with ulterior motives
- Adheres dogmatically to preexisting belief systems
- Makes authoritarian pronouncements of charismatic figures
- Abandons well-tested scientific hypotheses, with no contrary evidence
- Appeals to emotions
- Cannot be proven false by any conceivable test
- Does not apply to all occurrences

PREDICTION A forecast, based directly on the hypothesis, of some specific future occurrence that will happen if the hypothesis is correct, or explanation of a past, but not previously known, occurrence consistent with the hypothesis.

- Does not flow logically from the hypothesis
- Too general or vague to evaluate
- Allows too wide a margin for error

EXPERIMENTATION Objective sensing (seeing, hearing, etc.) of specific occurrences in physical reality for which predictions were made, capable of being reproduced by any suitably trained observer. (This list includes all observation flaws, since experimentation is, in effect, another observation.)

- Observers are not properly trained or equipped.
- Observers exaggerate or imagine phenomena.
- Positive instances are emphasized; negative ones are ignored.
- Unsupported personal anecdotes are relied upon as primary evidence.
- Measurements are subjective rather than objective.
- Observations are not reproducible.
- Con artists make fraudulent claims.
- Variables are not carefully controlled and/or carefully monitored.
- Results are not subjected to verification by other researchers.
- Human subjects modify their behavior because of knowledge of the hypothesis and/or experiment.

RECYCLING Does the experiment match the prediction? If yes, the hypothesis is supported, but not proven (a finite number of experiments cannot prove a general hypothesis; they can only support it). If no, the hypothesis must be modified or discarded.

- Naturalistic explanations of experiment are rejected.
- Dogmatic hypotheses are retained without openness to modification.
- Inference is drawn that an alleged cover-up automatically implies the truth of the hypothesis.
- Statistically insufficient or irrelevant data are cited to support a general hypothesis.
- Results that fail to support the hypothesis are discarded.

Here's a useful summary of the summary.

OBSERVATION *What, if anything, really happened?*

As accurately as possible, describe the observations that are supposedly consistent with an alleged power, experience, entity, or technique. Try to evaluate the extent to which these observations are reliable.

If they are unreliable, there is nothing to explain! If they are reliable, consider ways to explain them.

HYPOTHESIS *If something really happened, how can it be explained?*

Determine whether the explanation is falsifiable: Can it be shown to be false by any conceivable test? If the explanation is nonfalsifiable, it is not a scientific explanation.

Try to find related and relevant scientific hypotheses. If observations are consistent with these explanations, other explanations are not required.

If the observations cannot be explained in this manner, examine other explanations. Describe the power, experience, entity, or technique alleged to be consistent with them. Point out inconsistencies, if any, between this explanation and normally accepted ones. In the process, determine whether acceptance of the new explanation would require abandonment of any well-supported scientific hypotheses.

PREDICTION *What new observations can be expected if the explanation is correct?*

As accurately as possible, describe what would be expected to be observed under clearly specified, suitably controlled experimental conditions if the explanation is correct.

EXPERIMENT *What is actually observed?*

Observe what happens when these conditions are set up and the tests are carried out.

RECYCLING *How do the actual observations compare with the expected ones?*

Decide whether the experimental results match the ones that were predicted. To the extent that they do match, the new explanation is supported. To the extent that they do not, the new explanation must be modified or rejected entirely. When results are unclear, devise and carry out additional tests.

The Five Biggest Ideas of Pseudoscience

The five most widely believed ideas for which adherents claim scientific status are

1. UFOs and alien abductions
2. Paranormal OUT-OF-BODY EXPERIENCES such as astral projection and near-death experience and ENTITIES such as spirits and ghosts
3. Astrology
4. Creationism
5. Paranormal POWERS such as ESP and psychokinesis

In the following chapters, these ideas will be subjected to scientific scrutiny to see how well they fulfill the requirements of the scientific method. Each will be shown to be riddled with flaws that typify pseudoscientific thinking.

An Outline of Scientific Methodology

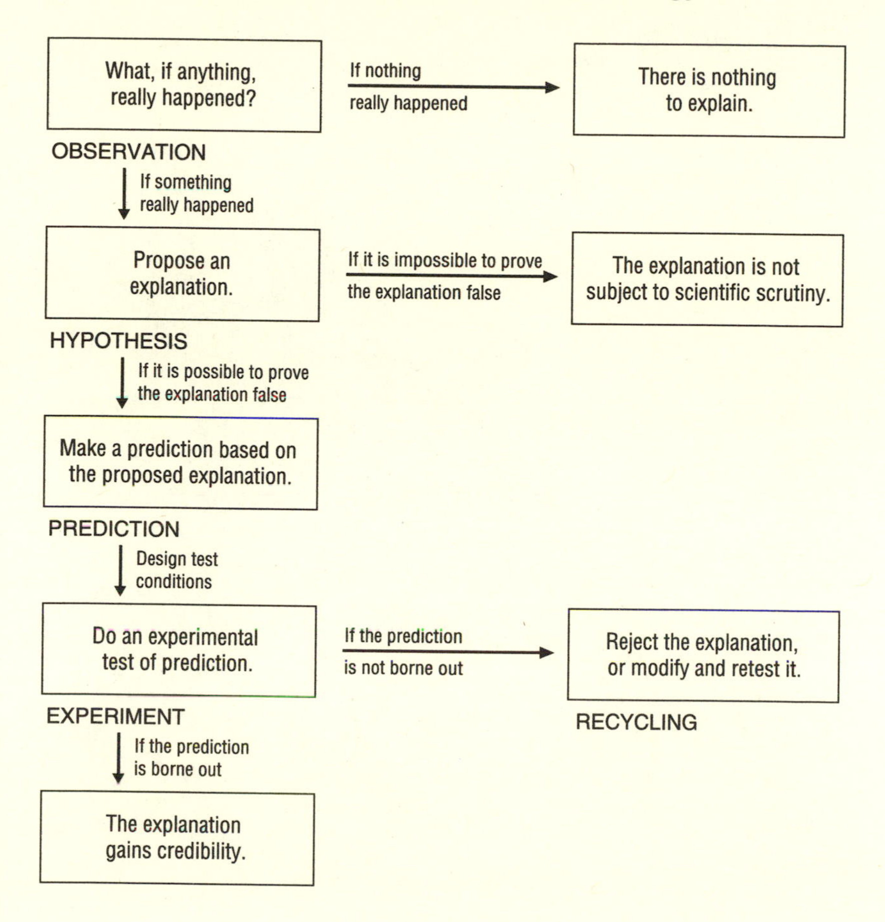

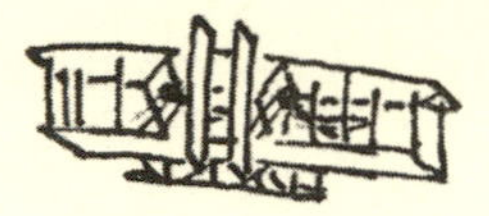

"LOOK — THERE GOES ONE OF THOSE UFOs AGAIN."

4 UFOs and the Extraterrestrial Life Hypothesis

My ideas caused people to reexamine Newtonian physics. It is inevitable that my own ideas will be reexamined and supplanted. If they are not, there will have been a gross failure somewhere.

A. Einstein

Do unidentified flying objects (UFOs) really exist? They most certainly do! No doubt about it: Many flying objects have not yet been identified. Have any formerly unidentified flying objects been convincingly shown to be alien spaceships? They most certainly have not.

It is ironic that *UFO* has become synonymous with alien spaceship, because, if an object had been identified as an alien spaceship, it would no longer be an "unidentified" flying object! Furthermore, sightings that turn out to be stray lights are not sightings of "objects." And although these lights may be moving, they are certainly not "flying." For these reasons, it has been suggested that *unidentified flying object (UFO)* be replaced by *unexplained aerial appearance (UAA)*, since the latter term does not put into people's minds preconceived notions that should not be there.

Early Observations: But I Saw It with My Own Eyes!

On June 24, 1947, nine unidentifiable moving objects in the sky were observed by Kenneth Arnold, a private pilot flying near the Cascade Mountains in Washington State. Mr. Arnold reported that the objects were flying "like a saucer skipping over water"—a flying saucer.

What happened shortly thereafter added to the mystery. Nine days later, in a New Mexico desert, an object was observed to fall out of the sky and land. Rumors began to circulate that the object might have been one of these flying saucers. The mystery was heightened five days later when the U.S. Army Air Force roped off the area.

Rumors began to spread that the bodies of four extraterrestrial beings (ETs) had been found in the wreckage. Rumors also spread that the government had sealed off the area so that it could remove the wreckage and bodies and cover up the existence of aliens. (For the record, the suspicious object that crashed in New Mexico in 1947 eventually proved to be a weather balloon.)

Other reports soon followed. In Roswell, New Mexico, in July 1947, Dan Wilmot and his wife saw a "big glowing object" that looked like "two inverted saucers faced mouth to mouth." Almost a week later, civil engineer Grady L. Barnett found "some sort of metallic, disk-shaped object" in the desert. The U.S. Army made the official announcement that the sighting and wreckage was a weather balloon. Many years later, the U.S. Navy and the CIA admitted testing high-altitude balloons for surveillance missions in this area at this time, hence the need for secrecy.

From the observation of an object falling from the sky and the government limiting access to the site, some people speculated that ***extraterrestrial vehicles and life-forms [had] reached the*** *surface of planet Earth.* The fact that governments have suppressed information, and even spread false information, does not mean they have suppressed information about ETs. The people who formulated this hypothesis took a quantum leap in the wrong direction. They wrongly drew inferences from supposed and nonrelevant observations.

Aviation Physiology

Identification of a flying object—or any other object—requires sufficient information about that object. Many objects appearing in the sky are viewed from a great distance, for a relatively short time, and only occasionally, thus making identification difficult if not impossible.

Even objects observed by large numbers of people for prolonged periods can be misjudged. For example, a full Moon appears larger near the horizon than when it is high in the sky. If you look at the Moon through a tube when it is near the horizon, it appears no larger than when it is overhead. This optical illusion is known as the "Moon illusion." One possible reason the Moon looks larger when it is near the horizon is that it is near objects we're used to seeing.

Objects observed in the sky by untrained observers are not necessarily what they appear to be. The ability to make accurate observations of phenomena in the sky, and to evaluate those observations properly, has to be acquired through rigorous training. Such training is especially important to airplane pilots. To reach their destination, pilots must be able to avoid collisions with other objects in the sky, as well as be able to compensate for visual illusions involved in gauging their altitude when approaching a runway during landing (depth perception). Pilots are therefore trained to be aware of the significant limitations inherent in their night as well as day vision and in how to compensate for those limitations.

To understand and appreciate some of the problems involved when a human identifies objects that appear in the sky, let's take a brief look at how a normal eye functions. The sense of sight is activated when light enters the pupil, a circular aperture in the center of the iris, then passes through the lens, and strikes the retina, a photosensitive layer at the back of the eye. This receptor records the image and transmits it through the optic nerve to the brain for interpretation.

The retina consists of light-sensitive cones and rods. Cones are concentrated around the center of the retina, and gradually diminish in number as the distance from the center increases. The small, notched area of the retina, located directly behind the lens, is called the fovea. This area contains a high concentration of cone receptors. Rods, on the other hand, are concentrated outside the fovea, and increase in number as the distance from the fovea increases. Because the rods are not located directly behind the pupil, they are responsible for much of our peripheral vision.

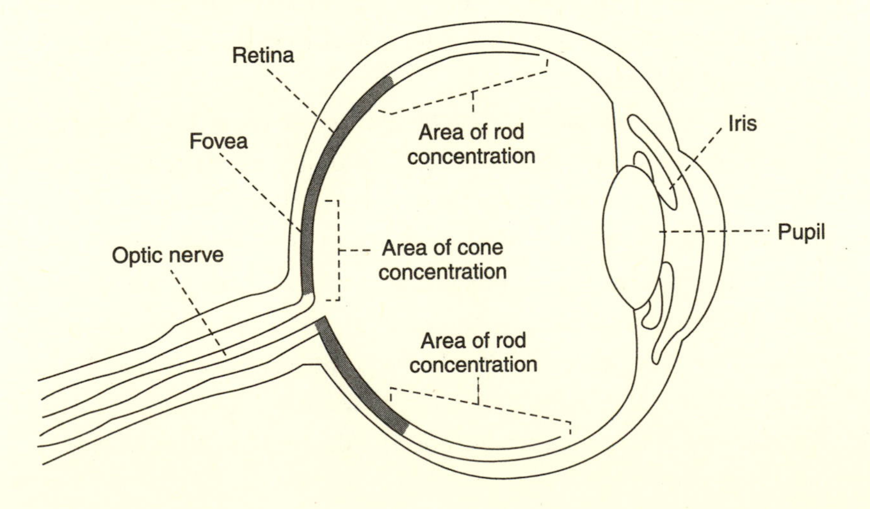

Although both cones and rods are light-sensitive, they serve different functions. Cones sense color and work best in bright light, whereas rods pick up only black and white and work best in low light. Since cones do not function well in darkness, night vision depends strongly on light picked up by rods.

The best vision in daylight is obtained by looking directly at an object so that the image is focused mainly on the fovea (cones). This tendency to look directly at an image does not serve a pilot as well at night. For this reason, pilots are trained to overcome this natural tendency. They are taught to expose more rods to the image by looking 5 to 10 degrees to the side of the object they want to see. Off-center scanning at night helps provide the visual acuity needed to avoid collisions.

Another problem with night vision is that although cones adapt quite rapidly to changes in light intensities, rods do not. Rods take up to 30 minutes to fully adapt to the dark. For this reason, pilots are trained to avoid bright lights for at least 30 minutes before a night flight. If they do encounter a bright light, they are trained to close one eye to keep it light sensitive, so they can see again once the light is gone.

Pilots are also trained to understand and avoid visual illusions, perceptions that differ from the way things really are. For example, after staring for a few moments at a single point of light against a dark background, such as a ground light or bright star, *the light will appear to move on its own.* A pilot who attempts to align the aircraft in sole relation to the light can lose control of the airplane. To guard against this illusion, pilots are trained to scan rather than stare at the sky.

In spite of all their training, pilots are still unable to completely eliminate the visual problems associated with flying. Their training, however, allows them to take preventive measures when appropriate, and goes a long way toward making their flights safer. Nevertheless, even pilots still occasionally mistake ordinary phenomena for extraordinary ones.

Hoaxes: Falsified UFO Hypotheses

In January 1967, Michigan teenagers Dan and Grant Jaroslaw said they saw a dark gray saucer hover over Lake St. Clair, then take off at high speed toward the southeast. They produced four photos of the disk. This incident received wide publicity, and the photos were studied by many experts. Nine years later, the brothers confessed that the pictures were a hoax: The UFO was actually a model suspended from a thread. Other hoaxers have used hot air balloons powered by candles to simulate UFOs. Hoaxes, however, do not play a major role in UFO reporting. Most witnesses are sincere. Misidentification on the part of a witness is far more likely than conscious fabrication.

Even after a hoax has been exposed, many UFO enthusiasts cling to their original belief and thus are not open to correction. One such group of enthusiasts had received purported photographs of UFOs from a man named Ed Walters. The group asked Rex and Carol Salisberry, two investigators who had earned the group's respect and confidence, to examine the photographs. When the Salisberrys reported to the group that Mr. Walters was adept at trick photography and had faked the photographs, *the group's response was to refuse to accept the report and to dismiss the Salisberrys.*

IFOs (Identified Flying Objects)

What is the true identity of items reported as UFOs? The most common one turns out to be the brightest of the planets, Venus.

Others include rocket launches, satellites in orbit, satellite debris (old rocket boosters, dead satellites, etc.) reentering the atmosphere, weather and other research balloons (especially up at 30,000–50,000 feet in the "jet stream" winds), and very high altitude military planes.

During a space walk at the *Skylab* space station, astronaut Ed Gibson remarked to fellow astronaut Bill Pogue, "Look over there; are those UFOs? There are hundreds of them." Pogue looked and saw "a cloud of metallic purple and violet sparkling objects that glistened with unusual sharpness and clarity." According to Pogue, he and Gibson became rather excited as they described these objects to Jerry Carr who was inside of *Skylab*. Carr turned down the lights inside and looked out the window, only to discover that what they were observing were shreds of aluminum-coated plastic that had been released when Pogue tore pieces from an aluminum-coated plastic blanket covering a piece of equipment he was repairing. In the twilight of space, these tiny reflectors had created a dazzling, twinkling cloud.

HUNDREDS OF UFOs

The J. Allen Hynek Center for UFO Studies (CUFOS) keeps close track of UFOs, and has on file ordinary explanations for 92 percent of all sightings. The balance could not be identified for lack of information.

Roswell Hysteria

Possibly the most publicized sighting was one that took place in July 1947, near the town of Roswell, New Mexico. In this instance, reports of the crash landing of an alien spacecraft were embellished with reports of the recovery of alien bodies. It was further rumored that the alien remains had been taken away by the U.S. Air Force and other alleged government co-conspirators for an autopsy. An alleged archival film of an autopsy being carried out on an extraterrestrial who was killed in the Roswell crash turned out to be a hoax.

One significant aspect of these reports is that, although the crash was originally reported in 1947, reports of alien bodies by the 1947 witnesses did not emerge until the late 1970s. By this time, reports of other alien life-form sightings were common, and could easily have stimulated the creative imagination of the witnesses.

Incidents that occurred near Roswell have been explained in a 1994 Air Force report, Roswell Report: Case Closed, as follows:

- The "unusual" military activities in the New Mexico desert were high-altitude research balloon launch and recovery operations.
- "Aliens" observed in the New Mexico desert were probably anthropomorphic test dummies that were carried aloft by U.S. Air Force high-altitude balloons for scientific research.

- Reports of military units that always seemed to arrive shortly after the crash of a flying saucer to retrieve the saucer and "crew" were actually accurate descriptions of Air Force personnel engaged in anthropomorphic dummy recovery operations.
- Claims of "alien bodies" at the Roswell Army Air Field hospital were most likely a combination of two separate incidents: (1) a 1956 KC-97 aircraft accident in which 11 Air Force members lost their lives, and (2) a 1959 manned balloon mishap in which two Air Force pilots were injured.

This report is based on thoroughly documented research supported by official records, technical reports, film footage, photographs, and interviews with individuals who were involved in these events.

Ancient Astronauts

There are people who claim that visitations to Earth by aliens have been going on for thousands of years and that such visits by "ancient astronauts" can explain many of the mysteries surrounding the cultural complexities and technological feats of ancient civilizations: How could ancient peoples have quarried, carved, hauled several miles, and erected huge stone statues on Easter Island? How were ancient peoples able to construct monumental Egyptian pyramids in an age devoid of the tools of modern technology?

The foremost proponent of this theory is Erich von Däniken, whose books have sold more than 40 million copies worldwide since 1970. Von Däniken claims, for example, that the stone carving on the lid of the funeral chamber of the Mayan King Pacal represents an ancient astronaut operating his spacecraft. The astronaut is alleged to be manipulating the spacecraft's controls with his hands and operating a pedal of some sort with his heel. On his nose is something said to resemble an oxygen mask. Outside the space vehicle there is said to be a small flame-like exhaust.

Those familiar with Mayan culture interpret the carving quite differently. The "controls" are representations of the Mayan Sun God in the background. The "pedal" is a sea shell, the Mayan symbol of death. The "mask" does not touch the nose. It is a piece of ornament worn by the king. The "flame" is the roots of a corn plant. Von Däniken's "astronaut" is actually the dead Mayan King Pacal!

Another claim of von Däniken is that markings or landing strips for a spacecraft designed by alien visitors are still visible in Peru. These long lines scraped in the desert are the so-called Nazca lines. The lines scratched in the desert, however, are much longer than airplane landing strips, and the soil there is much too sandy and soft to be used by airplanes. Such lines are much more likely to be worn-down pathways made by religious processions.

We will never know in complete detail how prehistoric peoples developed and applied their own arts and technologies. We do, however, know plausible mechanisms by which such feats could have been accomplished with the technology that was available to them.

Easter Island Statues

Descendants of the early natives have demonstrated a slow but plausible process by which the statuary can be created. Using old stone tools, it would take two teams of six men each about a year to complete a statue. Moving the statue over sand would take about 180 natives; moving it over harder ground would take about 90. At least 20 times this many people were available for this task. Erecting a statue could be accomplished by 12 men in 18 days, using log levers to raise the top of the statue a few inches and then inserting rocks under the raised part, until the statue was raised to a vertical position.

Egyptian Pyramids

Pyramid construction techniques evolved from techniques used to create simple mud-covered mounds and tombs, then brick ones, and finally ones made of stone blocks. Step pyramids evolved into the classic filled-in pyramid shape. The large blocks of relatively soft limestone could be quarried with hard stone tools. Logs imported from Lebanon and elsewhere could serve as huge rollers to help move the large blocks. Rafts could take the blocks up the Nile to the long sloping causeway leading from the banks of the Nile to the base of the Great Pyramid. Sledges of wood could be used to carry the stone blocks over land, and ramps of earth constructed to ease the blocks up to their position in the pyramid.

Is it possible for scientists to demonstrate that ancient astronauts have never been here? No. Proving a negative hypothesis is impossible. The burden, however, is on von Däniken to provide convincing proof that his astronauts did exist. This he clearly has not done.

Alien Abductions

In the 1950s, further complexity was added to the *extraterrestrial hypothesis that Earth is being visited by alien life-forms* when hundreds of people began to report that the alien life-forms had grown bolder. They reported that alien beings had kidnapped them, taken them aboard their flying saucers, and, in some cases, subjected them to painful medical examinations before setting them free.

Betty and Barney Hill may be considered the founding parents of the alien abduction movement. According to the Hills, in 1966, while driving in the White Mountains in New Hampshire, they were abducted by aliens, taken aboard a UFO, and then separated from each other. Betty said she was given a pregnancy test. Barney said a sample of his sperm was taken. Their story was brought to light only later, after they sought psychological help for recurring bad dreams. It was not until their psychiatrist, Benjamin Simon, put them under hypnosis that they reported the details of the incident.

The common abduction scenario involves aliens conducting "an ongoing genetic study" whose "focus is the production of children." Aliens descend from a spaceship while bathed in mysterious light. They enter closed windows or walls of homes at night. Their victim is alone, either awake or asleep. The subject is transported back to the spaceship through closed windows. Potential witnesses to these events are rendered unconscious. To make sure victims are not observed while in transit, they are rendered invisible.

After being taken to the spaceship's examining room, the victim's body is studied. The skin is minutely scrutinized from head to toe. Gynecological examinations are performed on females. Scraping and tissue samples may be taken from the genitals and other parts of the body. A small, round, seemingly metallic object is implanted in the victim's ear, nose, sinus cavity, and occasionally the penile shaft. (Post-abduction nosebleeds by victims are taken as evidence of nasal implantation.) Women may undergo egg harvesting, embryo implantation, or embryo extraction. Men may undergo sperm extraction.

Let us examine this ***extraterrestrial hypothesis that Earth is being visited by alien life-forms that abduct humans.***

One line of support given for its validity is that the stories told by many alleged abductees are quite similar. But this is not surprising given the frequent portrayals of UFO abductions in novels, movies, and television shows. Most people are already familiar with supposed details of these encounters, such as being manipulated in specific, strange, and often painful ways before being released, and experiencing "missing time" (not remembering what happened to them during a certain period).

Physical evidence to support the idea of pain infliction, such as scratches and cuts, is not necessarily the work of aliens. It is a common and ordinary experience to occasionally discover scratches and cuts with no memory of how they got there. And decreased awareness of time is also a common and ordinary experience, especially when people are anxious or under stress.

Whereas some people simply testify that they have been abducted, others provide strikingly detailed testimony while under hypnosis. The attempt to use hypnosis to remember details of a past event is called regressive hypnosis. But hypnosis is not truth serum. Not only can people fake hypnosis, they can even willfully lie while hypnotized. Although suggestions to recall details of a past event may produce increased recall, they can also cause people to include plausible details derived from unconscious fantasies, repressed sexual feelings, or memories from other times. Pseudomemories can include knowledge derived from other sources and incorporated into one's memory. Betty and Barney Hill, for example, may have incorporated imagery from then contemporary movies such as *Invaders From Mars* (1953) and TV programs about space aliens.

Another argument is that witnesses are able to pass lie detector tests to prove they are not just "making it all up." Although such tests are sometimes used in criminal investigations and elsewhere, their results are unreliable. They are not accepted in courts of law as definitive tests of truthfulness.

Nonfalsifiable UFO Hypotheses

A hypothesis is not scientific unless it is falsifiable (i.e., unless it is at least conceivable that evidence could be obtained that would disprove it); scientific hypotheses must lead to testable predictions. Pseudoscientific hypotheses often fail to meet this criterion. For example, in answer to the question, "Why don't bystanders in densely populated areas where alien abductions have been reported come forth as witnesses to these events?," believers in alien phenomena further complicate the ***extraterrestrial hypothesis (ETH)*** by claiming that ***witnesses' memories are erased by the aliens.*** This claim is nonfalsifiable. It does not lead to testable predictions.

Occam's Razor Applied to Overly Complex UFO Hypotheses

UFO hypotheses are more complex than observations warrant. In the absence of strong observational data, the invocation of extraterrestrial visitors or invisible Earth beings makes this hypothesis unjustifiably complicated. Although not all scientific ideas are simple, the complex ones are supported by repeatable evidence before they override the need for simplicity.

Not So Fast!

UFO hypotheses require abandonment of well-tested scientific hypotheses, with no contrary evidence. Unless aliens have exceedingly long lifetimes, travel to and from a hugely distant planet for the abductions would need to be accomplished at tremendous speeds, so quickly that the spacecraft might even have to travel

faster than the speed of light (186,000 miles per second). Faster-than-light speeds violate Einstein's well-tested ***theory of relativity*** and contravene the laws of physics as we know them.

Furthermore, space travel requires enormous energy expenditures. Einstein's ***special theory of relativity*** says that an object moving at nearly the speed of light becomes more massive, according to an observer not moving with the object. Because of the increased mass, more force is needed to accelerate the object. More force requires more fuel and thus increases the total mass, and so on. This means that a 10-person spaceship traveling to the nearest star in our galaxy at 70 percent of the speed of light would require millions of times the energy consumed by the United States in an entire year. Furthermore, the propulsion systems required for such journeys would have to be exceedingly more effective than any developed to date.

Another difficulty arises from the sudden changes in direction and motion attributed to UFOs, maneuvers not possible with known aircraft. Human beings can only withstand limited accelerations and still function. Some of the reported maneuvers would turn abducted humans to a soup-like substance, rendering them hardly suitable for examination, much less capable of return to Earth.

The SETI Project: Observations by Trained Observers

Even though scientific groups have found no evidence that extraterrestrials have visited Earth, the possibility remains that intelligent life exists elsewhere in the universe. An ongoing investigation of this possibility is called SETI (Search for Extra-Terrestrial Intelligence), begun by astronomer Frank Drake in 1960 and financed by the National Aeronautics and Space Administration.

The SETI project is based on the "testable prediction" that if civilizations do exist elsewhere in the cosmos, they might produce and emit radio wave patterns as a form of communication. SETI radio telescopes scan the skies 24 hours a day, searching for patterned radio wave signals that could indicate the presence of intelligence elsewhere in the universe. Currently, the search centers on nearby, old, yellow stars, and the frequency band searched covers the span from 1,000 to 3,000 megahertz, which contains typical frequencies for terrestrial radar systems. To date, no such patterns have been detected.

The Drake Equation: Is Anybody Listening?

How likely is the existence of extraterrestrial life-forms? From an astronomical perspective, relevant factors for their existence have been identified, and the overall mathematical probability of finding them has been estimated by astronomer Frank Drake in what is known as the Drake equation:

$$N = R_* \times f_p \times n_e \times f_l \times f_i \times f_c \times L$$

The symbols in this equation are defined as:

N Number of civilizations in the Milky Way with detectable radio emissions

R_* Rate of formation of stars that might support planets with intelligent life

f_p Fraction of these stars that have planets

n_e Number of planets per star system that have basic conditions necessary for life

f_l Fraction of n_e planets where life actually develops

f_i Fraction of f_l planets where intelligent life develops

f_c Fraction of f_i planets where technology advanced enough for space exploration develops

L Length of time the civilizations on the f_c planets last

Although the Drake equation was designed to be illustrative rather than precise, a number of astronomers have made the appropriate estimates and, as a result, think that other civilizations are highly probable. They argue that the universe is known to contain several billion galaxies, each of which contains billions of stars. Many of those stars are likely to have planets capable of supporting life. Evolution of intelligent and communicative life on such planets is certainly plausible.

A major difficulty is that the nearest one is at least hundreds of light-years away from us, which makes communication prohibitively slow. *The search for evidence of such civilizations, however, has thus far been unsuccessful.*

Alien Images

If alien life-forms did arrive on Earth, what might they look like? Naturally enough, we tend to suspect they would resemble humans. It is reasonable to assume that they would be more advanced technologically than we are, in order to overcome the formidable problems of space travel. It is assumed that these humanoids would be "us" as we will become in our distant evolutionary future. Their heads would be larger than ours (to house their larger, more intelligent brains) and their bodies would be relatively slight (because of dwindling physical activity, especially during space travel). This now standard image is the one usually depicted on the T-shirts and other paraphernalia sold in Roswell and elsewhere.

Contemporary images of alien life-forms differ significantly from previous ones. When the flying saucer craze began in 1947, aliens were described as little green men. These evolved into otherworldly beings bathed in light (1952), hairy dwarves (1954), goblins (1955), blobs (1958), 10-foot Cyclopses (1963), mothmen (1966), three-eyed giants (1970), insectoids (1973), robots (1977), reptilians (1978), fairies (1979), and lizard men (1983).

Well, either extraterrestrials have evolved at a mind-boggling rate since 1947, or they have been reinvented time and time again.

A Congenial Conclusion

The idea that we are not alone in the universe has been called a "congenial conclusion," a belief that would make life more interesting if true, and has an engaging air of plausibility. This idea is exhilarating to scientists as much as anyone. The prospect of a visit from intelligent extraterrestrials presents dazzling possibilities. Aside from scientific and technological interest, there is the hope that such visits might be of mutual benefit in understanding our place in the universe.

It would be a mistake to completely rule out the possibility, however slight, that we will be visited by ETs sometime in the future—or that we will someday be the visitors, the ETs, in some other solar system. To date, however, there is not a shred of credible evidence to support the belief that ETs have already visited us.

Extraordinary claims require extraordinary evidence. The extraordinary ***extraterrestrial hypothesis*** is not supported by ordinary evidence, much less extraordinary evidence. It is therefore untenable.

"ANOTHER ONE UNINHABITED. THAT'S THREE DOWN, AND SEVERAL HUNDRED BILLION TO GO."

"BUT YOU MUST ADMIT HALLUCINATIONS ARE MORE INTERESTING THAN DEPRESSION."

5 Out-of-Body Experiences and Entities

Science is the great antidote to the poison of enthusiasm and superstition.

Adam Smith

Spiritualism

Spiritualism deals with disincarnate entities or spirits. According to spiritualists, the spirit dwells in the physical body, but can leave it temporarily or permanently. When the body dies, the spirit departs from the body permanently and lives on in a world of disembodied spirits.

Let's examine three supposed manifestations of this phenomenon:

- Someone nearly died, and had a near-death experience in which the spirit temporarily left the person's body (an out-of-body experience).
- Someone did die, and now exists as a spirit that can communicate with the living directly as a ghost, or indirectly by

means of a "channeler" (also called a medium) who claims to have been invaded by the spirit and provides a channel through which the spirit can speak.

- The spirit of someone who is alive separates temporarily from the person's body and travels elsewhere (astral projection).

The Near-Death Experience

A patient is placed on the operating table and anesthetized, and then an operation is performed. Sometime during the operation, her heart stops beating, cutting off the supply of blood to her brain. She seems dead.

Death is the final frontier. Perhaps the most significant and profound question about human existence is, "What happens to her now?" Does she simply cease all existence, or does her existence somehow persist. Is there life after death? Immortality? One thing does seem clear: Living people desperately want to believe that they will not truly die. But, she's dead, so she cannot share any after-death experience with anyone else. Or can she?

Hospital staff attempt to use a defibrillation machine to resuscitate her. They succeed. Her heartbeat returns to normal. She is alive once again.

She reports that during the intervening period, she began to hear an uncomfortable noise, a loud ringing or buzzing. At the same time, she somehow left her physical body and was able to look down upon it as it lay on the operating table. She was a spectator to her own death. She continued to rise above her physical

> body as she entered a long, dark, spiraling tunnel and began an upward and peaceful journey. Her surroundings soon became brighter. She entered a radiant kingdom where she was illuminated by a distant, brilliant light. As she approached this light, she saw in her path a great godlike figure lit from behind—a being of light. She approached some sort of barrier or border, apparently representing the limit between earthly life and a next life. Sometime during her approach to the barrier and the figure, she realized that she had to leave this afterlife. The time of her permanent death had not yet come. She had to unite once again with her physical body. She awoke on the operating table in the hospital operating room.

Similar experiences have been reported by people who faced death as a result of an accident, life-threatening injury, or serious illness. They have been reported by men and women of all ages and cultures all over the world, not only by Judeo-Christians, but also by Hindus, Buddhists, and even skeptics. Near-death experiences of Indian swamis are substantially similar to those reported in the West.

What do these experiences mean? How might they be explained? What can we conclude from them?

Neurochemical Explanations of the Near-Death Experience

One possible explanation for the similarity of these experiences has to do with brain neurochemistry. It is a fact that an out-of-body experience, which is the sensation that one has left the physical body, can be induced with fair regularity, cross-culturally, by psychedelic drugs known to generate hallucinations and other distortions of perception by altering brain neurochemistry.

For example, dissociative anesthetics such as ketamines induce out-of-body experiences, and atropine and other belladonna alkaloids induce the illusion of flying. They have been ingested by European witches and American Indian shamans during religious ceremonies in which they experience religious ecstasy, soaring, and glorious flight. Also, hallucinogenic drugs, such as mescaline and LSD, are known to produce visions of striped tunnels and spiral chambers. The bright-light experience can result from stimulation of the central nervous system, which can mimic the effects of light on the retina. And, a point of bright light seen in otherwise dark surroundings creates a tunnel perspective.

Emotional and physical conditions capable of affecting the brain neurochemically in similar fashion can induce similar experiences. For example, in response to certain types of stress, the brain produces opiumlike substances called endorphins. These natural painkillers produce the same feeling of peace and well-being as that associated with near-death experiences. They are responsible for the natural euphoria of a "runner's high," in which the brain releases sufficient endorphins to counter pain and enable a long-distance runner to keep on going.

Such information is consistent with the hypothesis that ***the near-death experience is a neurochemical phenomenon.*** Supporting this hypothesis is the fact that general anesthetics given before and during operations affect the brain, not the site of the operation. Cardiac arrest also affects brain neurochemistry, by depriving the brain of its normal supply of blood.

It is possible that conditions such as cardiac arrest and anesthesia may, by altering brain neurochemistry, cause the brain to manufacture chemicals that create the near-death experience and brain states (hallucinations) that correspond to out-of-body experiences. Users (abusers) of hallucinogens often report seeing things when there is nothing to see, or they see things in ways that others do not. The fact that human biochemistry and reactions of the central nervous system to stimulation are universal can help explain the universality of these experiences.

The conditions under which this patient had her experience were far from controlled. To control the conditions, death could be induced artificially. Resuscitated patients could be asked to try to identify specific objects chosen through a double-blind procedure in which neither the patient nor the interviewer has knowledge of their identity. Artificial induction of death is, of course, highly unethical, and therefore such a study is both undesirable and unlikely.

On the other hand, a controlled study of whether someone can acquire knowledge of the physical world during an out-of-body experience might be conducted by bringing objects into the operating room after the patient has been completely anesthetized, and, if that patient reports a near-death experience, asking her to try to identify the objects.

The mysteries of the mind are many. Everything we know, sense, or feel is known or sensed or felt inside the neural network that is our brain. That brain can be fooled. Take, for example, the "phantom limb" phenomenon. People who have recently had a limb amputated continue to experience (in their brain) sensations (pain, etc.) that make it seem as if their limb were still intact. When someone says "it's all in your head," they may be correct. What seems to go on "out there" may actually be going on "up here."

Our brain can even create the perspective from which we "see" ourself. Imagine the last time you were lying on a beach. Describe what you see. Most people "see" themselves lying prone on a blanket. This memory, however, is not of a scene that was observed through their eyes. It is one that was constructed by their brain. In this partially constructed memory, the people appear to be "outside themselves." Such experiences are similar to that of the patient who reported she "somehow left her physical body and was able to look down upon it on the operating table."

Metaphysical Explanations of the Near-Death Experience

Another possible explanation of these experiences is the hypothesis that *there is a soul or psyche indwelling in the physical body. At death, this immaterial vital essence survives, leaves the body, and*

travels to another world (the soul hypothesis). A person who dies is said to have "given up the ghost."

One argument in support of this explanation is that people who have near-death experiences can often accurately report what was going on around them while they were clinically dead. For example, the patient may describe in great detail what members of the emergency room staff looked like and were saying while she was apparently dead. It is possible, however, that the information the patient supplied was obtained by ordinary means, namely, through her senses both before and during the procedure. The patient would likely have done a lot of reading and thinking about the impending operation. She may have become familiar with staff members during preoperation hospital visits and consultations and immediately before being anesthetized. Even under anesthesia, the senses are not completely turned off, especially the sense of hearing. In fact, the brain continues to function for a short time even after the heart ceases to beat. Hearing is the last sense to be lost, thus the patient may still be capable of hearing instructions given by the doctors and comments (even jokes) made by everyone in the operating room. Surgical patients recovering from anesthesia often recall auditory stimuli that were present during their surgery.

Another argument in support of the ***soul hypothesis*** is that patients' personalities are often transformed dramatically by the experience. They may lose their fear of death and gain a new sense of purpose in life. Nevertheless, the observation that something that causes a transformation seems real does not mean that it is truly real. Reading about completely fictitious characters in an inspiring novel can also produce transformations.

The transforming power of the near-death experience should not be so surprising in view of the patients' newly acquired belief that they encountered a godlike figure, were given a second chance, and were subsequently resurrected or brought back to life. The godlike figure, by the way, is seen by people of different religions as the God figure of their own religion, further indicating that the phenomenon occurs completely within the mind of the patient. Thus, it is possible that all the patients' knowledge about happenings during their experience was obtained by ordinary means.

The hypothesis that ***there is a soul or psyche indwelling in the physical body*** is more complex than the observations warrant. By adhering dogmatically to a preexisting belief system that posits souls, it responds in an emotionally appealing way to spiritual hungers and afterlife wishes. Thus it is not a conjecture of science. It is instead an article of faith. This being the case, science cannot accept this extraordinary ***soul hypothesis*** until extraordinary and compelling evidence is provided to support it.

What it is that is lost at death can be explained materialistically in terms of the various substances of which the body is constructed. Each human is a unique conglomeration of chemicals (molecules, etc.). We think, move, and feel because of information conveyed throughout the body by chemicals. At a fundamental level then, "life" may be viewed as "a chemical system that has a degree of complexity necessary to sustain life's vital characteristics (brain function, etc.)."

What is lost at death may not be an entity, but rather the particular arrangement of complex and interacting molecules that corresponds to life. If this arrangement is disturbed, the body sickens; if it is sufficiently upset, the body dies. Though its individual molecules remain intact for a short time, the body eventually disintegrates and its molecules are reabsorbed into various components of the environment.

Carl Sagan's Explanation: Born Again

The late Carl Sagan (and Stanislav Grof before him) endorsed the hypothesis that ***near-death experiences are personal recollections of birth experiences.*** Sagan's hypothesis predicts that people who were born by cesarean section and therefore did not have the "tunnel" experience at birth (the clinical stage of delivery in which the cervix is open and there is gradual propulsion through the birth canal), would not include a tunnel in their description of a near-death experience.

Experiments conducted by surveying people born by cesarean section and people born without the procedure, indicates that the hypothesis is false. People born by normal means were no more likely to report tunnel experiences than those who were born using cesarean section. Furthermore, studies of infant cognition indicate that, at the time of birth, brain development is insufficient for babies to remember specific details of the birth process.

Oxygen Deprivation?

Another hypothesis to consider is that ***near-death experiences are the result of neurochemical changes in the brain that are caused by the loss of oxygen in the brain when a patient's heart ceases to provide oxygenated blood to the brain.*** The hypothesis predicts that the blood of patients who report near-death experiences will not contain sufficient oxygen to maintain average brain function. This prediction is not borne out by experiment. A number of patients have had a near-death experience even though it was determined that their brain was not deprived of oxygen.

Spirits That Appear in the Form of Ghosts

Ghost stories are common worldwide. Listening to stories about ghosts while sitting around a campfire is a memorable part of many a camping experience. The ghosts in the stories are like the dancing flames that emerge as if by magic from the solid pieces of wood in the fire. These warm us temporarily before disappearing in the night air.

The term *ghost* describes the soul or specter of a dead person. This spirit is usually believed to inhabit another world from which it is capable of returning to the world of the living. Belief in ghosts is based on the notion that a human spirit is separable from the body and maintains its existence after the body's death.

Ghosts are said to be able to haunt certain locations, where they appear, displace objects, emit sounds such as laughter and screaming, ring bells, and even cause instruments to play. Noisy ghosts are known as poltergeists (literally, noisy spirits). They are the ones credited with certain malicious or disturbing phenomena such as throwing furniture or pots and pans around, making rapping sounds, and turning lights and electrical appliances on and off. They are also blamed for throwing stones and setting fire to clothing and furniture. Their activity is often said to concentrate on a particular member of a family, often a teenager. This individual is harassed when alone, but rarely when others are present.

What proof is presented for the existence of ghosts? Well, seeing is believing, so they say. Not long after the camera was invented, alleged photographs of ghosts began to appear. Probably the first such photograph was produced in 1862 by William H. Mummler of Boston, Massachusetts. Mummler said that when he developed a photograph taken of Mary Todd Lincoln all by herself, the photographic image included that of the spirit of her husband

Abraham Lincoln. (Lincoln died in 1865, so this would have been only a temporary departure from his physical body.)

Since then, thousands of seemingly inexplicable photographs of spirits have been produced. Many are quite crude and easily exposed as hoaxes. Images of spirits are superimposed onto the photograph or added to the original scene.

A classic example of this type of hoax involved Sir Arthur Conan Doyle, author of the famous Sherlock Holmes mysteries. Doyle heard and believed a tale told through photographs taken in Cottingley Glen, England, by two young girls named Frances Griffiths and Elsie Wright. Among the photographs was one showing Frances with four dancing fairies (miniature allegedly supernatural beings in the form of humans) in the foreground and another showing Elsie being offered a posy by a fairy.

Doyle "wanted" to believe in such supernatural creatures. His interest in spiritualism started as a hobby, but later became the focus of his life when his son was killed in World War I. He longed to communicate with this son and felt he could do so via the spirit world. Images of fairies served to confirm his belief in the existence of that world. In 1921, Doyle proclaimed his good news in a book, *The Coming of the Fairies.* Hundreds of people wrote to him describing fairies they had seen in their own gardens!

Although experts from Kodak in England could find no evidence in the negatives of superimposition of pictures of fairies, they did state that it would be possible to duplicate such photographs by this means. The possibility of trickery was dismissed because the girls were thought too young to perpetrate such a hoax and not knowledgeable enough in the use of photographic equipment. (It was later learned that Elsie had been employed at a photographer's shop where she specialized in retouching photographs.)

Much later, in 1978, Fred Gettings discovered that the four dancing fairy figures in the foreground of the photograph of Frances looked a great deal like the dancing figures found in a children's book called *Princess Mary's Gift Book,* published in 1915. The fairies could easily have been cardboard cutouts stuck in the ground.

Seeing/Hearing/Feeling Is Believing, Isn't It?

What about eyewitness reports of ghost sightings? After all, seeing is believing, isn't it? Well, eyewitness reports are notoriously unreliable, as any courtroom judge will testify. People commonly and unconsciously construct in their minds recollections that differ from actual events. They tend to fill visual and memory gaps with details to create a coherent picture. Eyewitness reports can also be unreliable because humans are prone to hallucinate under certain commonly occurring conditions, especially when they are somewhere between the states of complete wakefulness and sleep. Perceived images (ghosts, etc.) can occur suddenly and are not under voluntary control. They are very often vivid and realistic.

What about odd noises people hear coming from other parts of houses and apartments, creaks and taps, raps and bangs, doors seemingly opening and closing on their own? Naturalistic explanations abound for such phenomena: wood creaking as a result of expansion or contraction caused by temperature changes, sudden gusts of wind pushing against the surface of doors, tree branches brushing against walls and windows, and so forth.

And what about the "odd" or "creepy" feeling people get when entering certain places? Again, naturalistic explanations serve to explain these feelings: prior expectations, darkness filled in by imagination (fear of the dark), odd smells accompanied by unfamiliar dampness, and so on. These are all classic examples of pseudo-observations: The observers are not trained appropriately. They exaggerate or imagine phenomena. Personal anecdotes are relied upon as primary evidence. Observations are not reproducible.

Another such pseudo-observation is made when con artists make fraudulent reports. This was the case in Amityville, New York. Six members of the DeFeo family were murdered in a house there in 1974. In 1975, the house was purchased by George and Kathy Lutz. The Lutzs reported that, after they moved in, horrible and gruesome hauntings (house possessions by ghosts) occurred. They said these experiences were so bad that they decided to move out of the house after 28 days. A book about their experiences, *The Amityville Horror*, was a bestseller and was later made into a movie. Two years later, the entire story was revealed to be a hoax concocted by the Lutzes to make money.

Then there's the case of a 14-year-old girl, Tina Resch, of Columbus, Ohio. First, she saw the film *Poltergeist,* which depicted purported activities typical of active, noisy ghosts (poltergeists). Then, she reported similar phenomena in her own household. Although observers detected no mysterious movement of objects near Tina while she was being watched, as soon as they looked away, an object would fly across the room. Eventually, Tina was caught red-handed on videotape clandestinely throwing objects.

Haunted houses at Halloween may be great fun, but they also make evident our willingness to suspend our judgment—and our delight in being scared out of our wits.

Channeling: The Medium Is the Message

In addition to appearing as ghosts and communicating directly with the living, spirits are said to be able to communicate indirectly through channelers, individuals who are invaded by a spirit and provide a channel or medium through which it can speak. Through channelers, spirits are said to be able to impart wisdom, provide psychological counseling, and even make apocalyptic predictions.

Although channelers are found all over the world, they seem most likely to be found in California. There, J. Z. Knight, one of the best-known channelers in the world, channels a spirit named Ramtha. Ramtha is allegedly a 35,000-year-old warrior who swept through the [mythical] continent of Atlantis into India where he ascended into a higher consciousness and became a Hindu God-man.

His message, as channeled through Ms. Knight, is that God is a part of all human beings and that we create our own reality. Since we are all part of God and participate in the creation of reality, we are divine and therefore have no reason to feel guilt. Thus, we have within ourselves the means to achieve whatever goals we choose. To be happy, we simply have to create a happy reality ourselves. *It is the standard New Age message about self-created realities.*

This spirit guide has apparently acquired some remarkable abilities in addition to those he possessed 35,000 years ago. His advice on investing and other matters is often sought through Ms. Knight. This counsel does not come cheap. Ms. Knight (on Ramtha's behalf?) charges large fees. In addition, Ramtha books, videos, and cassette tapes are sources of advice—and royalties are payable to J. Z. Knight. The name *Ramtha* is copyrighted.

Ramtha predicted that, at the end of 1985, the United States would be engulfed in a major war, in 1988 a great holocaust would come and cities would be wiped out by disease, and a discovery in Turkey would reveal a great pyramid with a shaft reaching to the center of the Earth. None of this happened.

To be a successful channeler, one should be convincing, charismatic, intelligent, and well-read. Basically, the channeler's teachings are a mixture of Jungian philosophy, Western occult traditions, Hinduism, and contemporary positive-thinking attitudes. People buy into the claims and insights of channelers for many of the same reasons they endorse New Age philosophy in general: Through its emphasis on human potential psychology, it provides a sense of structure, discipline, and security. It also provides apparent confirmation of such beliefs by providing a direct encounter with an authority figure who sanctions those beliefs.

An important test of the hypothesis that ***channelers receive messages from disembodied entities*** would be to compare the accuracy of the information provided by the spirit with what actually happens. Unfortunately, the accuracy of this information is not readily evaluated because questions designed to solicit pertinent information are ignored by the spirit or are parried with evasive answers. Any predictions based on such answers are pseudopredictions, for the predicted occurrences may be too general to evaluate and may allow too wide a margin of error.

Another test of the channeling phenomenon is to analyze tape recordings of alleged spirits' voices to determine whether the speech pattern corresponds to the expected speech pattern of a person living at the time and place claimed by the spirit. For example, a spirit claiming to be a disincarnate thirteenth-century Scotsman from the Isle of Arran would be expected to sound like a thirteenth-century Scotsman.

Such was the claim of Lea Schultz of Lexington, Kentucky, who channeled "Samuel." Expert analysis by philologists (specialists in the field of language) concluded that Samuel's patterns were neither Scots English nor Scots Gaelic. In fact, they concluded that his speech patterns were "not those of a Scotsman of any century." In addition, they pointed out that the sounds of an authentic thirteenth-century dialect would be unintelligible to modern ears.

If channeling is not the result of input from disembodied entities, then what would explain this phenomenon? Some of the channelers may simply be putting on an act. They consciously pretend to receive messages. Some may be suffering from cryptomnesia, in which they remember something without recalling the memory's source, and so really believe that the thought is original or the result of extraordinary phenomena. Some may be in trance states in which autonomous parts of their consciousness may appear to be entities from "out there." And some may even be having delusions associated with mental illness.

Spirit Possession

A phenomenon related to channeling is "spirit possession," in which spirits "possess" or "inhabit" people. This phenomenon has been said to account for various complexities of human behavior.

For example, before people understood that epileptic "fits" have natural causes, these events were explained in terms of possession or seizure by spirits or demons (the word "epilepsy" comes from a Greek word that means "to possess or seize"). Epileptics were supposedly possessed by spirits that *threw them to the ground and tormented them with convulsions.* (The question,

"What possessed him to do a thing like that?" originally implied that a spirit or demon possessed the body and mind of whomever did "something like that.") Similar explanations have been given for certain forms of insanity, ecstatic trances, "speaking in tongues," visions, and oracular prophesies.

Possession by a bad spirit (a demon or witch) has been cited as the cause of some of the evil things people do. Possession by helpful spirits, on the other hand, such as the Holy Spirit or Spirit of Allah, has helped explain people doing good things and even accomplishing feats that otherwise would have been quite beyond their capability.

Techniques for depossessing one of a bad spirit (exorcising demons) and inducing possession by good spirits are still practiced throughout the world.

Reincarnation

Many people believe that one's spirit or soul has inhabited another being in a past existence and will be reborn in another body after death. This belief is known as reincarnation (also called transmigration). In primitive religions, this soul is frequently viewed as capable of leaving the body and being reborn, for example, as a bird, butterfly, or insect. The ancient Greek Orphists viewed the soul as being reincarnated in a human or other mammalian body. The Venda of southern Africa believe the soul of the deceased stays near the grave for a while before inhabiting another body.

This belief is held by millions of people around the world and is the doctrine of many religions. Asian religions, especially Hinduism, Jainism, Buddhism, and Sikhism, believe that reincarnation is affected by karma ("act"): What one does in this present life will have its effect in the next life.

One of the most famous stories about reincarnation is that of Bridey Murphy. According to this story, as told by amateur hypnotist Morey Bernstein in one of the bestselling books of 1956, *The Search for Bridey Murphy,* Virginia Tighe had lived a previous life in Ireland as Bridget ("Bridey") Kathleen Murphy.

Bernstein claimed to have uncovered this information in Pueblo, Colorado, when he "regressed" Tighe backward in time under hypnosis. In this state, Tighe spoke in an Irish brogue, sang Irish songs, and told Irish stories. She said she was born as Bridey Murphy in 1798 in Cork, Ireland, and provided details about her early childhood, her parents and brother, her marriage, her husband's profession, married life with her husband, and her funeral in 1864.

The story was well publicized and captured much attention. A Chicago newspaper obtained rights to republish parts of the book. In response, a rival Chicago newspaper decided to carefully check out the story. The paper discovered that Tighe had spent much of her youth in Chicago. As a child, she had heard stories about Ireland from an aunt who had been born there. And, across the street from Tighe's childhood home, there had lived another woman from Ireland who also told her stories about Ireland. That woman's name was Bridie Murphy Corkell!

Was the story a hoax perpetrated by Bernstein or Tighe? Probably not. What is more likely is that Tighe's "past-life memories" were only forgotten memories of childhood stories that she then wove together.

Astral Projection: Have Cord, Will Travel

It is alleged that the supposed separation of spirit from the body is not limited to the moment of death. Some people claim that under certain conditions their spirit can temporarily separate from their body (the tangible world) and travel elsewhere. This phenomenon is known as astral travel or astral projection. The entity that travels is referred to as an astral body. By some accounts, the astral body is a third component of humans, along with the material body and disincarnate soul.

Conditions under which astral travel is claimed to take place are yoga exercises, religious ecstasy, drowsiness preceding sleep, and hallucination. During the trip, the astral body floats effortlessly to some other place in the room (often near the ceiling), or much farther, since it is unrestrained by solid barriers.

Distance appears to be no problem when it comes to astral travel. Travel to other planets has been reported. Astral travel is likened to a physical waking state in which the astral body is able to look down upon and observe the world. A widely held belief is that the astral body remains attached to the material body by an infinitely elastic and very fine silver cord. Eventually, the astral body remerges with its material body.

Let's examine this hypothesis that ***an astral body indwelling in the material body can separate from it temporarily, travel to other planets, and make observations of the world.***

The hypothesis predicts that the astral body can give descriptions of places visited during its astral travels. Purported evidence of this ability is predominantly anecdotal. One test of the ability to travel astrally was provided in 1978 by a psychic named Ingo Swann. Swann claimed he had traveled to the planet Jupiter and as a result could give details about things not known to scientists. He provided 65 revelations, some quite specific. Later, *Mariner 10* and *Pioneer 10* spacecraft obtained information about Jupiter.

Swann's claimed observations were carefully compared with actual findings and information. According to an evaluation by

James Randi, 11 were correct, but the information was available in reference books, 1 was correct and not obtainable from reference books, 7 were correct but obvious, 5 were probable fact (scientific speculation), 9 were unverifiable because they were too vague, 30 were definitely incorrect, and 2 were probably incorrect. At best (giving him the benefit of the doubt), Swann's accuracy was an unimpressive and unconvincing 37 percent. In addition, travel by Swann's astral body to and from Jupiter in a few hours would have had to be accomplished at speeds greater than the speed of light.

A more prosaic hypothesis to explain the illusion of astral travel should be forthcoming when we better understand human neuroanatomy, the role played by belief in out-of-body experiences, and the interplay of prior knowledge and the imagination. A controlled test of this phenomenon would be to place an object in an inaccessible place, point out the location, and ask a person who claims the ability to astral travel to identify the object.

Immortality

So, do we simply cease to exist at the moment of death, or do we persist in some way after death? We are assuredly immortal in at least four ways:

- We live on in the genetic endowment we give to our offspring.
- We live on in the energy associated with our physical bodies that is released and ultimately partially conserved in other life-forms.
- We live on in the memories of people who have known or learned about us.
- And, we live on in the actions of others to whose moral and intellectual development we have contributed.

"MY ASTROLOGER SAYS ONE THING, MY GURU SAYS ANOTHER, MY PSYCHIATRIST SAYS SOMETHING ELSE — I DON'T KNOW WHO TO TURN TO ANYMORE."

6 The Astrology Hypothesis

The fault, dear Brutus, is not in the stars, but in ourselves.
William Shakespeare, *Julius Caesar,* I, ii, 134.

Reading the Entrails of a Newly Slaughtered Chicken

Since ancient times, people have resorted to a variety of pseudo-scientific interpretative techniques in their quest for ways of foretelling the future. Both curiosity and the advantages to be gained from foreknowledge of specific events have motivated people to seek such information.

Originally, these divination techniques were concerned with discovering the will of the gods, who were believed to control human affairs. This belief was the by-product of a fatalistic view of the human condition that said our fate was controlled by the gods, not ourselves. Later, other divination techniques were developed that dispensed with belief in a godly providence controlling human affairs.

Some of the early techniques were gory ones. Divination was

accomplished by "reading" the appearance and arrangement of the entrails of newly sacrificed animals such as chickens and sheep. Even living human beings were torn open and their entrails examined! Less gory techniques involved nondestructive observation of living things, for example, observation of the flight patterns of birds. If the birds flew to the left segment of a selected area of the sky, it was bad news. To the right, good news. Inanimate objects such as crystal balls, coins, and cards were also observed as well as manipulated.

Palmistry

Observation of humans has included observation of physical features of faces, heads, and hands. Hand or palm reading is still widely practiced. In palmistry (or chiromancy), the reading of character and the divination of an individual's future are derived from the lines, marks, and patterns on the palms of the hands.

Everyone is born with a unique set of fingerprints and a unique palm print (pattern of lines on the palm). These patterns are genetically determined. Certain of these lines appear consistently. They are identified by palmists as head lines, heart lines, life lines, and fate or destiny lines, and are alleged to be useful in predicting a person's future.

Even the length and shape of individual fingers are said to provide insights, for example, the thumb or Venus finger will provide the palmist with information about how stubborn and how giving you are. Hand size, shape, color, texture, fleshy lumps, the depth of lines, and even the manner in which the hands are held when being read are taken into account. Magnifying glasses may be used to look for minute details.

Palmists are able to use clues gained from hand features such as

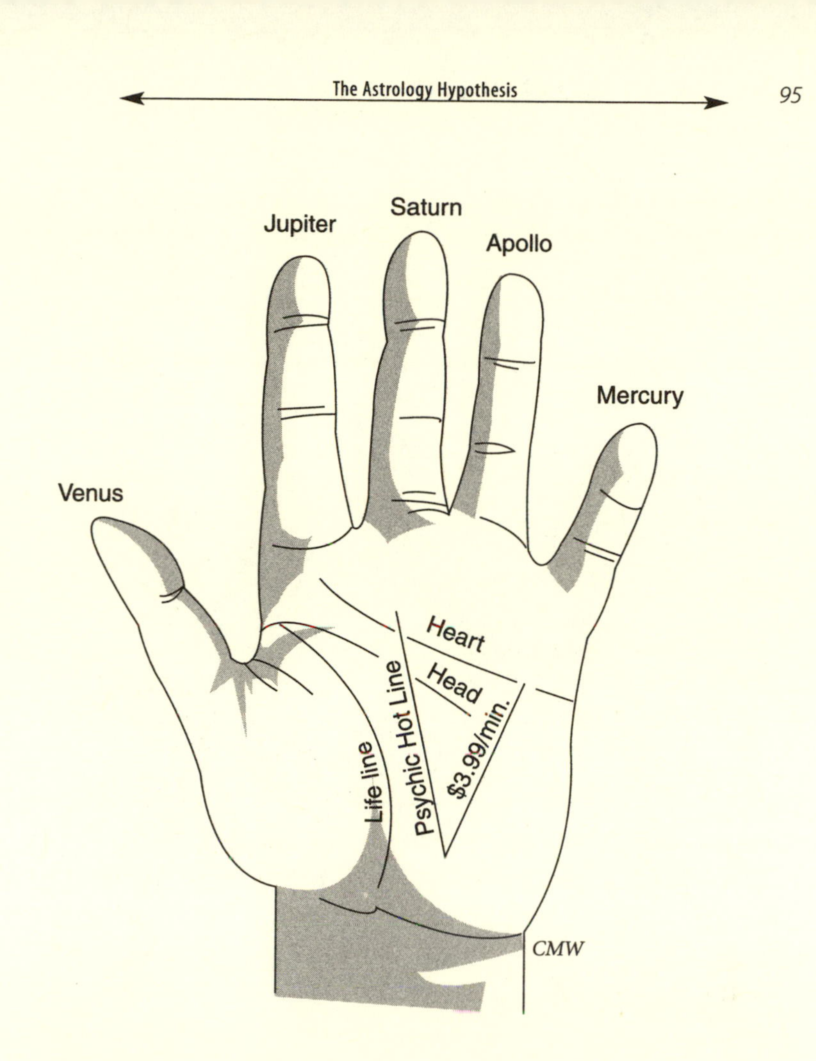

calluses, bitten nails, and swollen finger joints to "read" a person and thereby astound the unsophisticated! "You handle a lot of heavy or rough material every day." (Certain occupations roughen the skin and leave it callused.) "You've been worried about something lately." (Nervousness results in nail-biting.) "Your doctor has informed you that you have arthritis." (Hands are routinely examined in medical diagnosis to provide evidence of arthritis (swollen joints) as well as circulatory problems (nails that do not readily regain their color after being pressed).

"YOU ARE GENEROUS. YOU LIKE TO TRAVEL. YOU HAVE GARISH TASTE IN CLOTHING."

Numerology

What's in a name? According to numerologists, much more than you'd suppose. Take, for example, THOMAS MARTIN. Using the following "figure alphabet," in which the letters in each column are assigned values corresponding to the number on top of the column, numerologists translate each letter into a number, and then add these numbers.

1	2	3	4	5	6	7	8	9
A	B	C	D	E	F	G	H	I
J	K	L	M	N	O	P	Q	R
S	T	U	V	W	X	Y	Z	

THOMAS translates as 2 + 8 + 6 + 4 + 1 + 1 = 22; MARTIN as 4 + 1 + 9 + 2 + 9 + 5 = 30. THOMAS MARTIN thus translates into 22 + 30 = 52.

In the simplest as well as most popular numerological method, the figures in Thomas Martin's total are then added: 5 + 2 = 7. This sum happens to be a primary number (i.e., a number from 1 to 9) so the process is complete. If the sum is not a primary number, the figures are added together until the result is a primary number. For example, a sum of 29 becomes 2 + 9 = 11; then, 11 becomes 1 + 1 = 2.

According to numerologists, the name number "7" has "grand possibilities." Thomas Martin can bend his natural talents to art, science, and philosophy, often attaining greatness in a chosen field. Although he may rise to dramatic heights, he should remember that his success will depend upon quiet planning and may often require deep meditation. And so forth.

More complex methods include additional input such as birthdate numbers. As with other pseudoscientific fortune-telling techniques, people will ignore the parts that don't fit them, and focus on those that do seem accurate (or fit their image of how they'd like to be). If the forecast is vague enough, it can serve as a one-size-fits-all prediction.

Numerology has no scientific basis. There is no plausible explanation of how one's fate can be predetermined by one's name and birthdate.

Graphology

Another technique said to provide insight about humans is graphology, the analysis of character through handwriting, which was studied in ancient times and given a psychic significance. As a field of study, it is said to be older than the Pyramids of Egypt. Graphoanalysis involves scrutiny of small details and comparisons of various styles. Many employers hire graphologists to analyze these aspects of handwriting samples provided by prospective employees, to determine their suitability for a position.

Some graphologists claim that monitoring one's handwriting can provide a useful health indicator. Others maintain they can eliminate a person's "bad" habits by improving the person's handwriting. Investigations into the validity of handwriting as a character assessment tool have failed overwhelmingly to show any significant positive correlations. The small number that have evidenced such correlations appear to be based on information revealed in handwriting samples having autobiographical content.

Scrying

Scrying is divination that involves staring into a reflective object such as a mirror, a still pool or bowl of water, or a crystal ball. These objects are intended to help open one's awareness to psychic insight. Crystal gazing seeks visions allegedly seen in a ball of rock crystal, preferably quartz. It can involve elaborate rituals for cleaning the ball and for conducting the crystal-gazing sessions. The globe is said to mist up from within before presenting its visions of the past, present, and future.

"HOW DO YOU WANT IT—THE CRYSTAL MUMBO-JUMBO OR STATISTICAL PROBABILITY?"

The Ouija Board

Another alleged way to communicate indirectly with spirits is by using a device called a Ouija board. The word ouija is a combination of the French and German words for yes. The board began as a game used for entertainment in the nineteenth century, but evolved into a metaphysical pursuit, a supposed "portal to the spirit realm where one may find the answers to life's many mysteries."

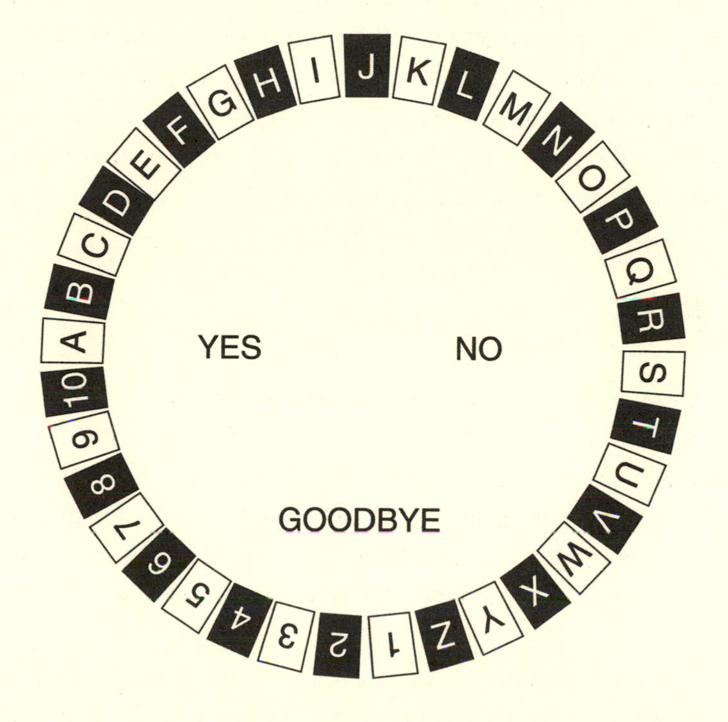

The Ouija board has all 26 letters of the alphabet and the numbers from 0 to 9 drawn on it. It can also have the words yes, no, and goodbye. A triangular pointer (or planchette) rests on the board. One or two people at a time may use the board. They rest their hands lightly on the pointer. When two people use the board, one is on each side, and both hands of each are rested on the pointer. They ask the Ouija a question. In response, the pointer allegedly moves in such a way as to spell out or point to a spiritualistic and telepathic answer on the board. When the Ouija decides to conclude a session, it allegedly directs the pointer to goodbye.

Pseudoscientists assert that dead spirits move the pointer to communicate messages from the spirit world. In reality, however, unless someone is deliberately moving the pointer to spell out a message, small subconscious movements in the hand are responsible for the movement. In this case, the Ouija operator has no conscious awareness of having moved the pointer and may be genuinely surprised at the answer.

I Ching

Manipulation of collections of objects includes I Ching, the ancient Chinese practice of fortune-telling or foretelling the future by examining the pattern that results after coins or yarrow (a perennial plant) stalks have been tossed. Guidance provided by the I Ching is said to depend to a large degree on perceptive interpretation of the resulting patterns in light of the questions being posed.

The more complex yarrow stalk method involves tossing 50 yarrow stalks that are then divided into bundles of yarrow stalks by an oracle who then "calculates" the lines. The much simpler coin method works as follows: Three coins are tossed. The number of heads versus tails corresponds to either a broken or an unbroken line. The first throw determines the bottom line of a six-line pattern called a hexagram. Five subsequent throws complete the set of six broken or unbroken lines. Sixty-four different combinations of six broken or unbroken lines produce sixty-four different hexagrams. Each has a symbolic name signifying a different condition of life. If properly understood and interpreted, they are said to contain profound meanings applicable to daily life. Those meanings are expressed in the form of answers to questions such as, "What does the future hold for me?"

The I Ching, or Book of Changes, is based on the idea that the universe is made up of two equal and complementary forces, Yin and Yang. Yin can be the passive, female cosmic principle, Yang the active, masculine cosmic principle. Yin can stand for dark, while Yang stands for light. Since, according to this idea, everything is made from Yin and Yang, differences between things are the result of different proportions of Yin and Yang.

Yin and Yang are translated in the hexagrams as broken lines (Yin) and unbroken lines (Yang). Each hexagram is made up of two groups of three lines or trigrams. When two trigrams come together, they will stand in varying degrees of accord or dissent with each other. If they are in accord, the hexagram signifies something good, pleasant, or fortunate. If they are in discord, it signifies something bad, unpleasant, or unlucky.

Here are two hexagrams:

The one on the left is said to stand for peace, harmony, and balance because the three Yang lines of the lower trigram provide the strongest possible support for the three Yin lines in the upper one. The one on the right is said to stand for stagnation, and is most unfavorable because the Yang trigram is bearing down with all its weight on a passive, yielding Yin trigram.

Tarot Cards

Divination or fortune-telling using tarot cards involves examining the pattern or sequence of cards produced after a tarot deck is shuffled by the person seeking advice and handed to a fortune-teller. The fortune-teller lays out a few of the cards (either selected at random by the questioner or dealt off the top of the shuffled deck) in a special pattern called a spread. Card meanings depend on whether or not they are upside down, their position in the spread, and the meaning of adjacent cards.

The Roman Catholic church condemned tarot reading as the Devil's invention in the fourteenth and fifteenth centuries. It was banned in several European cities, and tarot cards were burned in the marketplace in Nuremberg. Still, tarot reading survives; the tarot card divination industry continues to cater to people in search of easy answers and glimpses into the future.

The standard tarot deck consists of 78 cards divided into two groups, the Major Arcana, which has 22 cards (also known as trumps), and the Minor Arcana, which has 56 cards. Twenty-one of the cards of the Major Arcana are numbered from I through XXI; the Fool card is unnumbered. Major Arcana cards or tarots include the Magician (I), the Hanged Man (XII), Last Judgment (XX), the Moon (XVII), and the Devil (XV). Cards of the Major Arcana refer to spiritual matters and important trends in the questioner's life.

Minor Arcana cards comprise kings (or lords), queens (or ladies), knaves (or servants), and knights of wands (agriculture), swords (warriors), cups (clergy), and coins or pentacles (commerce). Cards of the Minor Arcana deal mainly with business matters and career ambitions (wands), love (cups), conflict (swords), and money and material comfort (coins). The four suits in modern-day decks of 52 cards are present in the tarot deck as wands (clubs), swords (spades), cups (hearts), and coins or pentacles (diamonds). Values in each suit progress from ace to ten, then page (knave or jack), knight, queen, and king. Elimination of knights leaves the 52 members of modern decks of playing cards.

Astrology: Reading the Entrails of the Heavens

The most widely practiced form of divination is astrology, as in

> I was born when the Sun was in Gemini.
> Wouldn't you just know he's a Pisces!
> Is Mercury in his Aries too?

Although the central hypothesis of astrology is rarely stated explicitly, one possible version is that *the positions and move-*

ments of particular celestial bodies at the moment of a human being's birth predetermine that individual's personality and other characteristics, and that these celestial bodies influence day-to-day events during one's lifetime. Besides individual readings, astrology is also used to forecast the destinies of collective entities, such as companies, groups, or even entire nations.

The predetermined personality characteristics described by astrology are quite similar to the characteristics associated with the Greek gods for whom the planets were originally named. This connection is not surprising considering astrology's earliest roots.

Astrology and astronomy were both developed by the Babylonians about 3,000 years ago. They were the first astronomers to keep systematic and precise details of when and where the planets (wandering stars, to them) appeared in the sky. Their study of these records and charts led them to conclude that planets moved in predictable patterns. As astrologers, they drew conclusions about the relationships between what they considered lucky and unlucky days and the alignment of the stars on those days.

From this information, the Babylonians developed techniques for reading the future from the predictable movements of the planets. Later, the Greeks combined Babylonian techniques with Egyptian ones and added a few of their own. These were summarized and published as the *Tetrabiblos* by Claudius Ptolemy around 150 CE. This work is the standard reference for today's astrologers.

Astrology spread to India in its Babylonian form. Islamic culture absorbed it as part of the Greek heritage. Western Europe was in turn strongly affected by Islamic astrology. Then, when Western astrology became known in China through Arabic influences in Mongol times, the Western version was integrated into preexisting Chinese ideas about an intelligible cosmic order. In the later centuries of Imperial China, the idea became so firmly rooted in Chinese culture that a horoscope (astrological forecast) was made whenever a child was born and also at important junctures in the child's life. Even today in China, cesarean sections are scheduled to coincide with "lucky" horoscope days.

Sun Sign Astrology

All forms of astrology are based on celestial objects. Sun Sign Astrology, the most common one, is based on the zodiac, 12 segments of the sky, each named after a constellation (a configuration of specific stars) that was in its region about 2,000 years ago, when Ptolemy wrote the *Tetrabiblos.* These zones are the ones commonly identified in the personal horoscopes that appear daily in over 1,000 newspapers in North America.

Not only is the Babylonian division of the zodiac into 12 zones a completely arbitrary invention (Egyptians, for example, classified the Sun's path into 36 divisions), but the constellations themselves are nothing more than apparent star groupings named by the ancients to honor people, animals, or a significant object (e.g., Libra the Balance) in their mythology.

The stars in each constellation are not even close together in space. They are at widely varying distances from each other and from Earth but happen to lie along similar sight lines. For example, the three stars that make up Orion's Belt, Alnitak, Alnilam, and Mintaka, appear to be very near each other, but actually lie 815, 1,345, and 920 light-years from Earth, respectively.

Of major significance in astrology is the Sun sign, the particular zone of the zodiac in which the Sun was located at the moment of one's birth. According to astrologers, the exact moment is required because the Sun sign changes during a particular day.

The Sun sign is considered of primary importance because it is thought to be the most powerful of all celestial influences upon an individual. It is said that this aspect colors the personality so strongly that an "amazingly" accurate picture can be given of the individual who was born when the Sun was "exercising its power" through the influence of a certain astrological sign.

Here's a list of the signs of the zodiac and the corresponding dates when the Sun entered them in ancient times.

Aries the Ram—March 21
Taurus the Bull—April 20
Gemini the Twins—May 21
Cancer the Crab—June 22
Leo the Lion—July 23
Virgo the Virgin—August 23
Libra the Balance—September 23
Scorpio the Scorpion—October 24
Sagittarius the Archer—November 23
Capricorn the Goat—December 22
Aquarius the Water Bearer—January 20
Pisces the Fishes—February 19

Although the natal charts or horoscopes that appear in daily newspapers and popular magazines are based solely on Sun signs, other celestial aspects, such as the exact location of one's natal Sun and the position and movement of the Moon as well as planets, are included in more sophisticated versions.

For example, while your Sun (which determines character) may have been in Gemini when you were born, your Moon (which rules emotions) might have been in Aries, the planet Mercury (ruling the mind) could have been in Scorpio, Mars (ruling your

"THE REASON HE'S NEVER SEEN A CONSTELLATION IS HE'S CONVINCED THERE REALLY ARE WHITE LINES CONNECTING THE STARS."

speech and movements) could have been in Taurus, while Venus might have been in Capricorn, giving you an essentially Capricorn attitude in romantic, artistic, and creative matters.

Another consideration is your ascendant, the sign rising on the eastern horizon at your moment of birth. This factor is said to modify your personal appearance and also help form your true inner nature. The ascendant and Moon sign are considered the two most important positions in any natal chart, after the Sun sign.

In addition, each sign has its own house, a specific zone of the sky fixed with respect to the horizon. The first house is usually defined as that sector of the sky immediately beneath the eastern horizon and contains those planets that will rise within approximately the next 2 hours. The second house contains objects that will rise 2 hours later, and so on. The influence on these houses by other signs is also of significance in astrology.

Thus your total personality is a blending of Sun, Moon, and ascendant influences, as well as planetary, house, and other effects. Interpreting a chart involves consulting astrology manuals that give standard interpretations of all these influences.

The Stars Impel but Do Not Compel

A common claim by astrologers is that a detailed natal chart is able to indicate tendencies to honesty or dishonesty, cruelty, violence, fears, phobias, and even psychic ability. It can also indicate inclinations toward or away from drug addiction, promiscuity, frigidity, homosexuality, multiple marriages, a disturbed childhood, alienation from or neurotic attachments to relatives, hidden talents, and career and financial potential. Also revealed are susceptibility and immunity to accidents and disease and secret attitudes toward drink, sex, work, religion, children, and romance. In other words, according to astrologers, *no secrets are hidden from the accurately calculated natal chart.*

Insights obtained from people's horoscopes are said to help them develop to their fullest potential while avoiding potential pitfalls that may have been revealed. Insights obtained from studying another person's horoscope can help make one more understanding and tolerant of the deeply ingrained characteristics with which that person was born. For example, an Aquarian won't seem as rude when she roots into your private life once you realize she was created with an uncontrollable urge to investigate people's motives.

Let's now examine the claim that astrology is based on mathematical data and astronomical information and is a full-fledged science. We'll see how and why this claim is flawed, thus placing astrology in the realm of pseudoscience.

Observation Flaws

The original information used to construct the manuals that guide astrologers in their interpretation was obtained by people who had an incorrect and incomplete view of the physical universe. Their beliefs incorrectly placed the Earth at the center of the universe, the universe they described contained far fewer celestial bodies than are now known, and the paths of the bodies they described were incorrectly believed to be overlapping circles.

In addition to requiring knowledge of the relative positions of celestial bodies, astrological observations require knowledge of the precise time those bodies were in particular positions. According to the Code of Ethics of the American Federation of Astrologists, an "opinion cannot be honestly rendered unless based on a horoscope cast for the year, month, day, and time of day plus geographic location of the place of birth." If this is so, then the data originally used to construct the astrological charts upon which horoscopes are now based are unacceptable because they do not meet this standard: accurate instruments for telling time became available only in recent centuries, long after the original charts were made.

Hypothesis Flaws

Many astrologers adhere dogmatically to preexisting belief systems. At the time astrology was first proposed, our planet was thought to be the center of the universe. Since then, as a result of centuries of application of the scientific method, astronomy discarded that perspective. Further, several additional planets (Uranus, Neptune, and Pluto) and many planetary moons have been discovered since the early days of astrology. "Astrological effects" of these bodies on people's personalities are ignored by all but the most diligent astrologers.

In addition, during the past 2,000 years, the Earth has shifted its axis of rotation to such a degree that the signs of the zodiac have slipped westward by about 30 degrees from the original positions described in the *Tetrabiblos.* Corrections for this shift have not been made in astrological calculations. In other words, *the zodiacal constellations named in ancient times no longer correspond to the segments of the zodiac represented by their signs.* Four thousand years ago, the Sun was in the constellation of Taurus on the day of the spring equinox (March 21); 2,000 years later it was in Aries; today it is in Pisces. The shifting of the axis of rotation affects not only Sun signs, but all other aspects of the natal chart: Moon sign, planetary signs, ascendant, and sign influences on houses.

No explanation is provided for the link between the Greek gods' personalities, planet names, and individual human traits. Further, what is crucial about the moment of birth? Is this defined as the first time the child's head appears, and does it depend on the time of labor? What about cesarean section? Would moment of conception be better? What about other initial conditions: mother's health, various aspects of the place of delivery? And what about artificial insemination—or the possibility of cloning humans?

Astrology abandons well-tested scientific hypotheses, with no reason or evidence for doing so. One such hypothesis is biology's theory of genetics, which says that personality traits can be explained at least in part in terms of inheritance of appropriate genes from prior generations. Biologists are currently in the process of mapping the structure of the molecule that encodes such traits, DNA, and engaging in lively discussion about the interplay between genetics and environmental (not celestial!) influences in determining personality traits.

No Known Mechanism for Conveying Astrological "Influences"

Physics has found only four forces in nature: gravity, electromagnetism, weak nuclear, and strong nuclear. Of these, the two nuclear forces have zero strength outside the nucleus, and the electromagnetic force is interfered with or blocked by the presence of matter of various kinds. This leaves only gravity as a potential source of "astrological" (celestial) effects.

Let's see how well gravity holds up as a candidate for "influencing" humans at their moment of birth. The closest celestial object to Earth is the Moon. Undoubtedly, the Moon exerts a significant influence on planet Earth: Lunar gravitational forces cause the tides. Since tides result from the Moon's gravitational attraction on the oceans, and humans are mostly water (about 70 percent), some astrologers argue that the Moon must also exert an influence on the water in humans. Undoubtedly it does. The relevant question, however, is not *does* the Moon have a gravitational attraction to the water in humans. The question is *how much* of an effect does it have, and how does this effect influence a person's personality while that person is being born?

The Moon causes tides only in large unbounded bodies of water, such as the world's oceans. Even lakes, unless they are exceptionally large, are negligibly influenced. Furthermore, the well-established law of universal gravitation states that every mass exerts a gravitational force of attraction on all other masses in the universe, and the greater the distance between two objects, the smaller the gravitational force between them. When distances and masses are taken into account, calculations show that the person who helps deliver a baby exerts a greater gravitational "influence" on the baby than does the Moon. Planets, which are orders of magnitude more distant than the Moon, produce even less gravitational force.

If the force of gravity is not a viable candidate for "influencing" humans, could there be a yet-to-be-discovered force that is? Astrologers point out that electrical and magnetic forces remained undiscovered until the nineteenth century. And the two nuclear forces weren't discovered until the twentieth century.

Yes, it certainly is possible that an as-yet-to-be-discovered force does exist. But, until that force is detected, its existence remains pure conjecture and cannot be cited in support of the astrology hypothesis. There is no obvious reason why celestial bodies and their movements should influence the human condition. Since no one has offered a plausible explanation for how celestial bodies exercise this effect, the effect must be considered extraordinary. As such, science requires extraordinary proof if this effect is to be accepted. In the absence of such proof, astrology's hypothesis must be rejected.

Astrology's hypothesis is also unacceptable because it violates the scientific standard of falsifiability; it is phrased in such a way that it cannot be proven false by any conceivable test. Once astrologers argue that birth charts indicate only what a person will potentially become, the statement cannot be disproved because it adjusts to fit all data, even contradictory data. When the predicted influence of celestial bodies does not evidence itself in correspon-

"YOU WANT PROOF? I'LL GIVE YOU PROOF!"

ding personality characteristics, astrologers can say that the potential for those characteristics exists, but it has not been expressed.

Astrologers want to have it both ways. If they predict that "a stranger's advice will get you into trouble," and a stranger's advice does get you into trouble, they're right. If you heed this warning, and avoid such trouble, you have your horoscope to thank for it!

Why do so many people cling to this ancient belief in spite of its many defects? Ignorance of the flaws is one reason. Many people are still unaware, for example, that it is the Earth that revolves around the Sun, and not the other way around (as the ancient astrologers believed).

But ignorance is not the only reason for clinging to this belief. Astrology has a strong emotional appeal. Naturally enough, we are all interested in learning whatever we can about ourselves.

Astrology is appealing because it purports to provide this information in an immediate and reliable way. It provides a belief system to help satisfy human spiritual hungers and gives advice on how to improve health, avoid difficulties, and even find the right mate.

At least some of the people involved in providing this information have ulterior motives. Given the volume of newspaper astrology columns, books, articles, mail-order computer-horoscope merchants, radio and TV spots and talk shows, hot lines, personal consultations, as well as sale of charms, greeting cards, T-shirts, and the like, astrology is a multibillion-dollar business. Certainly, many astrologers are being enriched, and handsomely. Some writers of astrology columns don't even pretend their advice is accurate. At the bottom of their column they write, "For entertainment purposes only."

Prediction Flaws

If astrological predictions flowed from a valid hypothesis, and there were universally accepted methods of calculation, it would be expected that predictions made in different daily newspapers on the same date and for the same Sun sign would be substantially in agreement. They are not. One horoscope says today is a good day for taking risks, another urges extreme caution; and so on.

Most of these predictions are too general or vague to evaluate anyway. Suppose an astrologer predicts that sometime during the next week you will be disappointed by the actions of someone close to you. This prediction is so vague that it will "fit" a wide variety of likely experiences. Unless you spend the next week in total isolation, it's hard to conceive how the prediction would not seem correct.

French psychologist Michel Gauquelin conducted an experiment to determine whether people would be able to reject a horoscope that did not correspond to their date of birth. His ploy: A computerized astrology profile for a particular individual was sent to people with unrelated birth circumstances. The person whose horoscope was sent was a notorious mass murderer, executed in 1946 for murdering 27 people and disposing of their bodies in a tub of quicklime located in a secret chamber of his home. The mass murderer's horoscope said, in part: "instinctive warmth or power is allied with the resources of the intellect, lucidity, wit," "endowed with a moral sense which is comforting," "a tendency to be more pleasant in one's own home."

The people who received the mass murderer's horoscope were told the profile corresponded to their own date of birth and were asked to evaluate its accuracy. Ninety-four percent of the 150 people who replied said they were portrayed accurately in the horoscope. Ninety percent of their friends and family shared this assessment.

The responses can be understood in terms of the well-established Forer effect: When presented with a long list of general and specific personality characteristics that supposedly apply to them, people tend to accept traits they desire to have, and ignore the rest. This is also known as the Barnum effect after P. T. Barnum, the circus impresario, who said that a good circus has a "little something for everybody."

Experiment Flaws

In astrology, personal anecdotes are often relied upon as primary evidence. When a good friend of yours who has followed her horoscope for years tells you that she firmly believes it has consistently been both accurate and beneficial, you should keep in mind that such assertions must be evaluated objectively and in their entirety. Human minds are highly susceptible to the power of suggestion and will therefore believe things about themselves that others do not. Human subjects will modify their behavior because of knowledge of the hypothesis and/or experiment. For example, studies have shown that people who are aware of the identifying characteristics predicted by their astrological sign are more likely to claim those characteristics than those who are not aware of them.

Recycling Flaws

Astrology's hypothesis is retained without openness to modification or to being discarded entirely. It is dogmatic. Regardless of the fact that objective tests have repeatedly found that predictions made by astrologers are not borne out by experiment, people refuse to modify or reject this hypothesis.

In one such test, half of a group of subjects were given birth charts compiled for their date of birth and asked to rate how accurately the charts fit them. The other half of the test group were given birth charts as nearly opposite as possible to their correct birth charts and asked to rate how accurately the charts fit them. The results were virtually identical. Opposite birth charts were judged to be no less accurate than "real" ones.

A Sign of the Times

Does the fact that the astrology hypothesis is accepted by billions of people and has survived for thousands of years have any bearing on the "truth" of astrology? No. The validity of scientific hypotheses is based on the scientific method, not popular approval. Truth in science is not a matter of which hypothesis gets the most votes, it is a matter of which hypothesis makes predictions that match experimental evidence.

In 1975, a letter was written alerting the public to the fact that there is no evidence for the claims of astrology. Prime movers behind the statement were Bart Bok, an astronomer of international standing, and Paul Kurtz, professor of philosophy at the State University of New York at Buffalo. They circulated the statement, mostly among members of the National Academy of Sciences, and published it with 186 signatures, many from Nobel Prize winners. They proclaimed:

> We, the undersigned—astronomers, astrophysicists, and scientists in other fields—wish to caution the public against the unquestioning acceptance of the predictions and advice given privately and publicly by astrologers. Those who wish to believe in astrology should realize that there is no scientific foundation for its tenets. . . . It is simply a mistake to imagine that the forces exerted by stars and planets at the moment of birth can in any way shape our futures. Neither is it true that the position of distant heavenly bodies make certain days or periods more favorable to particular kinds of action, or that the sign under which one was born determines one's compatibility or incompatibility with other people.

Would you have signed it? We would.

DARWIN AT A TURNING POINT
ON THE OTHER HAND, IT WOULD BE A LOT LESS TROUBLE TO GO ALONG WITH CREATIONISM.
S. Harris

7 The Creationism Hypothesis

In science the important thing is to modify and change one's ideas as science advances.

Claude Bernard

If you wanted to learn about the activities of Lord Ernest Rutherford, the New Zealand physicist who discovered in 1911 that atoms have nuclei, you would have at your disposal abundant information about Rutherford and his work (detailed photographs, actual equipment used, etc.).

If you wanted to learn about activities that took place much earlier, for example, the life of Democritus, the Greek philosopher who lived from about 460 to 370 BCE and introduced the idea of atoms, you would have a much more difficult task. You could review existing accounts of ancient Greece, examine a small number of artifacts, and try to locate any as-yet-undiscovered artifacts or records.

Now, suppose you wanted to go back even further in time—to the earliest moments of the universe itself. What records or artifacts could you use to examine the lifetime of the universe?

On a clear night, you can focus on one of billions and billions of artifacts that provide information about the history of the universe: a star. What's significant here is that you are seeing the star not as it is, but as it was. Stars are so distant from Earth that it takes years for their light to reach us. The light from the nearest star in our Milky Way galaxy (collection of stars), Alpha Centauri, took four years to reach our solar system. Thus, we are seeing it as it looked four years ago. *When we see stars, we are looking into the history of the universe.*

In the years since the light that we now see left its star, the star might have expanded, contracted, or even exploded (as a supernova). Light from Andromeda, the nearest galaxy to the Milky Way galaxy, takes about 2 million years to reach us. Out beyond the galaxies, there are celestial objects called quasars, whose light has traveled at least 10 billion years to get here. *The clear indication here is that the universe must have been in existence for at least 10 billion years.*

Galaxies are grouped with other galaxies in clusters. After observing hundreds of clusters of galaxies, astronomers have determined that every known (observed) cluster of galaxies in the universe is moving away from every other cluster of galaxies. In other words, *the universe is expanding.*

Since the universe is currently expanding, it is reasonable to hypothesize that, at an earlier time, clusters of galaxies must have been closer together. Carried to its extreme, this hypothesis suggests that at one time all the matter of the universe was compacted together. Because it is known how far the clusters are from each other today and approximately how fast they are moving away from each other, it is possible to estimate that this single compacted unit existed about 12 to 15 billion years ago and has been expanding ever since.

Evolution of the Universe

As the expansion began, the early universe must have been extraordinarily hot and dense because its entire mass was in an extremely compressed state. In the ***big bang theory***, astronomers hypothesize that this primeval fireball expanded incredibly rapidly, creating space as it did so. Entities that we recognize as living things could not have existed under the extreme conditions of this stage in the history of the universe. Thus, although the universe has a 12-billion to 15-billion-year history, life does not. Living things could not exist until the appropriate atoms were formed and the universe had expanded to a point when the density and temperature somewhere in the universe became low enough for the chemistry of life to become possible.

In astronomy's ***big bang*** scenario, the expansion of the universe reached a stage about 4.5 billion years ago when the material that was to become planet Earth was part of a gaseous cloud called a nebula. As this nebula began to rotate, most of its matter gradually contracted into the center, and eventually became our Sun. Smaller accumulations became planets. The third aggregation of mass near the Sun became planet Earth. During the nebula and planet formation stages, conditions were still not compatible with the existence of living things.

Evolution of Living Things

According to biology's ***theory of evolution***, less than 1 billion years after the formation of Earth, about 3.8 billion years ago, conditions finally became suitable for the appearance of the first entity having the characteristics of life (3.5 billion-year-old rocks have yielded fossils of primitive microorganisms). *These findings indicate that all living organisms have a common ancestry.*

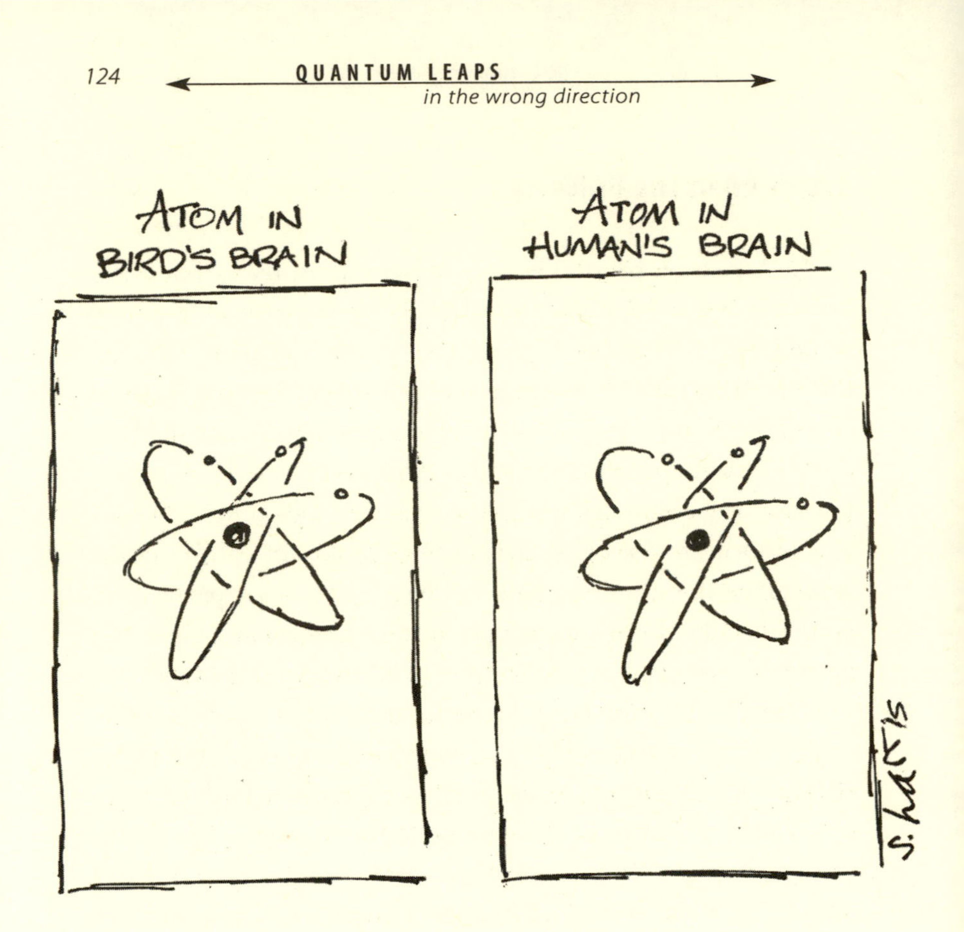

In time, simple one-celled organisms evolved into more complex one-celled organisms, more complex one-celled organisms evolved into simple multicelled organisms, and simple multicelled organisms evolved into more complex multicelled organisms. Eventually, species (interbreeding groups of similar organisms) evolved. Ultimately, the most complex of species, modern human beings, evolved.

Just as astronomy's big bang scenario is supported by artifacts such as stars of widely varying longevity, biology's theory of evolution is supported by artifacts of life such as fossil remains whose longevity has been established experimentally by radiometric dating. For example, these techniques have established that approximately 30,000 species of shelled marine animals called brachiopods existed about 500 million years ago. Brachiopods number approximately 300 species today.

The theoretical mechanisms by which evolutionary change takes place have been abundantly supported by experiment. Fundamental to these mechanisms are mutations, changes in the genetic material that determines anatomical as well as biochemical characteristics. Such changes have been observed repeatedly in nature and have also been induced in the laboratory. Genetic mutations result in variations that accumulate in populations (organisms belonging to the same species or interbreeding group) so that every population contains an immense amount of genetic variability.

It has been observed in natural populations of plants and animals that some genetic variants are more successful in surviving and reproducing than others. If populations acquire differences that prevent them from mating with each other, they can eventually become different species. Evolution of such reproductive barriers has been observed both in the wild and in experimental situations.

Evolution of Planet Earth

According to geology's ***plate tectonics theory***, while life was evolving on Earth, the overall structure of the planet was also evolving. The "plates" in plate tectonics are the giant mobile slabs into which the outermost layer of the present-day four-layered structure of Earth is broken. Tectonic refers to structural deformation in the Earth's outermost layer.

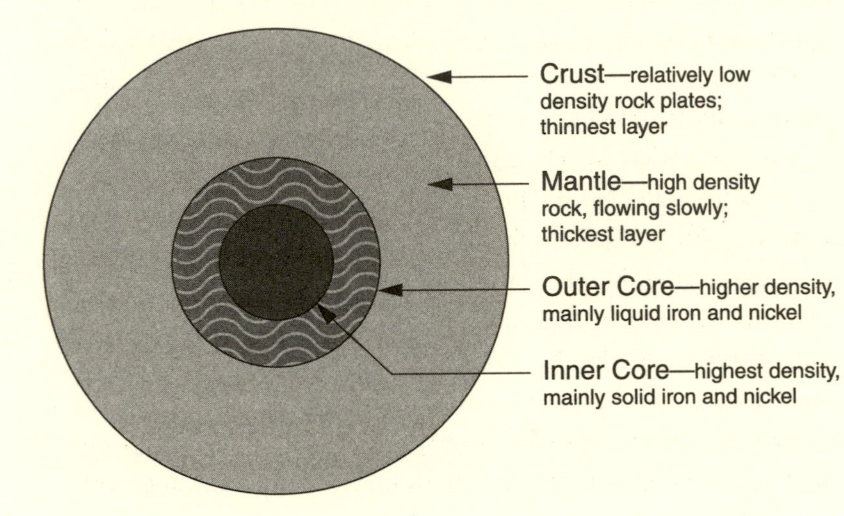

Geology's theory includes a comprehensive, detailed scenario for the evolution of Earth's layered structure as well as features that exist on and in its outermost layer. Artifacts that help geologists construct the scenario include fossils and rocks whose ages have been determined experimentally and whose placements provide clues to their history.

Warping the Loom: The Tapestry of Science

These three theories, biology's ***theory of the evolution of living things*** (introduced in the 1860s), astronomy's ***big bang theory of the evolution of the universe*** (introduced in the 1920s), and geology's ***plate tectonics theory of the evolution of Earth*** (introduced in the 1960s), are woven together in a rich tapestry of theories about the life and times of the universe. Each theory is interconnected to and supported by the others. Indeed, one of the hallmarks of science is the convergence of information obtained by its various disciplines.

For example, astronomers have estimated that planet Earth is about 4 billion years old. This estimate is based on measurement of the relative amounts of the elements hydrogen and helium in our Sun, which is a star. (A star's helium abundance indicates how long it has been converting its original fuel, hydrogen, into helium. Earth formed at roughly the same time as the Sun, and so, their ages must be comparable.) The age of Earth estimated by astronomers is the same as that estimated by geologists from measurements of plate movements and by biologists from measurements of coral growth.*

The "Scientific" Quick Creationism Theory

There is an idea that predates the ***big bang/plate tectonics/biological evolution*** theory of the sciences by several thousand years. It is

---

*Additional evidence for biology's ***theory of the evolution of living things,*** astronomy's ***big bang theory of the evolution of the universe,*** and geology's ***plate tectonics theory of the evolution of Earth*** is available in our previous book, *The Five Biggest Ideas in Science.*

supernatural creation, a religious idea woven out of a strict literal interpretation of Genesis, the first book of the Bible. It presents a significantly different scenario from that of the sciences.

In this scenario, (1) the universe is created in all its complexity by the command of God in six days of 24 hours each, no more than 6,000 to 10,000 years ago, (2) all the species that exist or have existed were created at the same time, (3) Noah's flood 4,500 years ago was universal and accounts for the deposition of rock strata and fossils, and (4) humanity was dispersed into many races and tongues at the Tower of Babel. This two-millennia-old idea of faith, in which God created a complex universe in just six days, has been called ***quick creationism.***

As scientific insights into the nature of the universe were generated during the past 2,000 years, many religious people realized that a strict literal interpretation of Genesis was no longer appropriate, and they rewove their tapestry accordingly. The six days of creation described by Genesis were reinterpreted as activities that had taken place over billions of years. These people replaced the earlier theory with ***gradual creationism***, a tapestry based upon religious faith, but tempered by scientific insights. In an evolutionary sense, they evolved from adherence to a scenario in which God creates a complex universe in just six days, to a scenario in which God's creative involvement is in the context of a 12-billion to 15-billion-year period.

One group of religious people has not evolved in this sense. This group is unwavering in its strict literal interpretation of the Bible. To support their belief, they claim that ***quick creationism*** is supported by scientific evidence. They claim that the Bible is a book of science as well as a book of religion.

This augmented theory, commonly called ***scientific creationism*** or ***creation science***, will be referred to here as ***"scientific"***

quick creationism to avoid confusing it with scientifically tempered versions of ***gradual creationism.***

Once these "scientific" quick creationists contend that their theory is one of scientific knowledge as well as religious faith, it opens the theory to evaluation according to scientific standards. Let us therefore apply scientific standards to claims that ***"scientific" quick creationism theory*** is a scientific theory comparable or even superior to ***big bang/plate tectonics/biological evolution theory.***

The "Worldwide Flood" Claim

Geologists have discovered that fossil remains are distributed in layers in which the oldest fossils are generally located in the lowest layers and the youngest ones in the uppermost ones. Any exceptions to this order are readily explained by geologists as the result of subsequent deformation (folding or upheaval) of the layers. Geologists attribute these findings to a multimillion-year process involving sequential deposition of layers of sediment containing the remains of organisms (the earliest or oldest ones being deposited first, etc.).

Such a scenario would be impossible if the universe was created only 6,000 to 10,000 years ago. Nevertheless, ***"scientific" quick creationism theory*** explains these discoveries in terms of a worldwide flood described in Genesis. According to this scenario, torrential rains caused the flooding of the entire Earth surface for 371 days. Prior to this event, God commanded Noah to build a boat large enough to accommodate his family, thousands of species of animals, and about 1 million species of insects, along with the food required to sustain them. The boat had to be sturdy enough to withstand 40 days and nights of rain. Only those living things that found refuge in Noah's Ark survived the flood.

When the rains ended and the waters subsided, animals not aboard the ark were buried in sequence. "Scientific" quick creationists claim that animals that lived in the sea would be buried first. Next to be buried would be the slow-moving amphibians and reptiles, then faster animals, and finally humans. This order corresponds to the order found in the vertical geologic columns unearthed by geologists.

Science Responds to the "Worldwide Flood" Claim

Two major questions must be answered "yes" for the worldwide flood claim to be true: Was it technically possible for an ark to be built in ancient times that would hold and sustain at least two of all species? Was there a worldwide flood in the first place?

An answer to the first question begins with calculation of the size of the ark described in Genesis. Translating the cubit dimensions of the Bible into feet results in an ark that is 450 feet long, 75 feet wide, and 45 feet high. According to ship architects, this exceeds by 150 feet the maximum length of a seaworthy wooden ship. If the length exceeds 300 feet, unavoidable warping and stresses will cause the hull to leak so much that the vessel will sink.

Even if a seaworthy 450 foot long, wooden ship could be built, the technical problems involved in getting more than a million species of animals and plants aboard, then housed and fed for 371 days would be insurmountable. Even if, to save space, only the eggs of those species that reproduce by eggs (dinosaurs, reptiles, fish, amphibians, and birds) were brought aboard, almost none of these has an incubation period longer than the 371 days they were afloat. Most of the eggs would have hatched while still on board the ark. The orphaned hatchlings would require a great deal of constant care and would therefore place a tremendous burden on people caring for them. This task (and numerous others) would have to be accomplished by just eight people: Noah and his wife, their three sons and the sons' wives. To further complicate matters, the collective gene pool of these eight humans would have to account for all of the racial diversity and physical distinctions now found in the human race!

Space would also be a serious problem. Items on board would include: a 371-day supply of stored food, cages for animals as large as dinosaurs, freshwater tanks and saltwater tanks (the rapid change in salinity during a flood would kill nearly all fish not aboard the ark), waste products (a single elephant can produce 40 tons of manure in a year), plants in soil, as well as living space for the eight humans.

Some plants could be taken aboard as seeds, but many plants do not reproduce by seed, so they would have to be taken as adult plants. Someone would have had to be able to gather these plants and animals from all over the world. Parasites and infectious microorganisms that cannot survive outside of their host animals or humans would have to travel "aboard" these hosts without destroying them. Noah and his family would have had to be infected with the likes of syphilis, smallpox, and leprosy for over a year.

To those who take this story literally, discovery of remains of Noah's large and sturdy ark would be evidence of the worldwide flood from which it was to provide refuge. Legend has it that Noah's Ark came to rest on top of Mount Ararat in Turkey, and it is there that people have searched for the evidence. No such evidence has been found. Since 1829, explorers have searched in vain on Mount Ararat. Claims to have seen and photographed the ark have never withstood careful scrutiny. Samples of wood alleged to come from a 5,000-year-old ark have been found to date from about 800 CE.

The answer to the second question is even more critical. Technical questions about arks become moot in the absence of a worldwide flood. Flood stories are common to many cultures, and some predate the one given in Genesis. These include hero equivalents of Noah (Zinsuddu in a Sumerian flood story, ca. 3,000 BCE, Utnapishtim in a subsequent Babylonian text, the Gilgamesh Epic, Xisuthus in a succeeding Babylonian version, etc.), a god secretly warning the hero of an impending flood that would destroy humankind, an ark landing on a mountain, and even the sending out of birds until one finds dry land and does not return.

The similarity of the Genesis version to previous tales does not necessarily mean that the Genesis version should be discounted as folklore rather than fact. The real test of a worldwide flood scenario is to deduce the kinds of physical evidence such a scenario predicts, and see whether those predictions are borne out.

Genuine floods leave physical evidence in the form of sedimentary deposits in a narrow band at the same level (same date of occurrence). This leads to the prediction that a worldwide flood would leave a worldwide band of sediment at the same level in geographic columns of sediment in all areas that had been dry land before the flood. The absence of such a band of sediment means that the prediction of this hypothesis is not matched by experimental evidence.

Another problem with the worldwide flood scenario is the location of the quantity of water required to flood Earth to a degree that 17,000 foot high mountains could be submerged completely. By far, the largest potential source of terrestrial water is the ice in the North and South Polar regions. Even if this entire supply of ice melted, the level of the oceans would rise no more than 30 feet. Thus the Earth contained insufficient water to produce the required depth of water for the flood.

Might the water supply have descended from outer space? A comet containing the necessary amount of frozen water would be so massive that its impact would literally shatter the Earth. Even if the required amount of water was somehow made available, where would it all go to allow the flood to subside?

Have major floods occurred in the past? Strong evidence suggests that they have. However, they were local, not worldwide, events. They may have seemed worldwide to people whose world was limited by geographic constraints. When these stories of past local floods were shared by travelers, they could have assumed that the floods had occurred at the same time, and blended them into a worldwide phenomenon.

The "Dinosaurs and Humans Were Contemporaries" Claim

According to the ***theory of evolution***, species were not created all at once. They appeared at different times; for example, dinosaurs first appeared 250 million years ago during the Mesozoic Era and died out about 65 million years ago, whereas humans did not appear until the Cenozoic Era about 200,000 years ago. If dinosaurs died out long before humans evolved, then evidence that the two coexisted would be a serious blow to evolutionary timetables. Therefore, according to "scientific" quick creationists, the finding of evidence that footprints of dinosaurs and humans were made at the same time on what was once a muddy riverbank near Glen Rose, Texas, is proof that the ***theory of evolution*** is seriously flawed.

s. harris
MESOZOIC ERA
CENOZOIC ERA

Science Responds to the Claim That Dinosaurs and Humans Were Contemporaries

In 1986, paleontologist Glen Kuban studied these famous footprints. He was able to determine that what looked like human footprints were actually those of a three-toed dinosaur. Erosion had blurred the image, but careful examination revealed evidence of the three toes. Furthermore, the spacing between the allegedly human footprints was significantly different from that typical of a human.

The "Second Law of Thermodynamics Is Contravened" Claim

According to the *second law of thermodynamics*, a fundamental and well-tested scientific law, whatever order exists in a closed system (one in which no matter or energy enters or leaves) will eventually run down and revert to a disorganized, random state: ordered states inevitably become less orderly.

According to the *theory of evolution*, the evolution of living things is the incremental production of *more* organized states (living things) from *less* organized ones (molecules, etc.). Thus, claim "scientific" quick creationists, since planet Earth is a closed system, evolution (more ordered states evolving from less ordered ones) is not possible.

Scientists Respond to the Claim That the Second Law of Thermodynamics Is Contravened

The *second law of thermodynamics* is not violated by evolution because planet Earth is not a closed system. It receives significant amounts of energy from outside its system (from outside the planet), namely from sunlight. Energy received from the Sun makes it possible for living things to develop. For example, during photosynthesis, a process in which energy from the Sun is absorbed with the aid of chlorophyll molecules, order increases when water and carbon dioxide molecules end up as more highly structured sugar molecules. After the organism dies and can no longer utilize this energy, the process is reversed and the organism decomposes.

The "Evolution Has Never Been Observed" Claim

"Scientific" quick creationists contend that evolution has never been observed directly, and therefore ask why anyone should believe that evolution occurred in the way biologists suggest.

Scientists Respond to the Claim That Evolution Has Never Been Observed

The problem here is a misunderstanding of what biologists mean by evolution. Biologists define evolution as a change over time in the gene pool (collective genes) of organisms belonging to the same species and occupying a particular geographic area. Such changes have been observed. One example is insects developing a resistance to pesticides over the period of a few years. The collective effect of similar changes has made it possible to generate the diversity of all living things, including new species.

Evolution has also been observed retrospectively in the sense that its predictions about what would be expected to be discovered in the fossil record—geographical distribution of species, and so forth—have been borne out.

The "Random Chance Cannot Explain It All" Claim

"Scientific" quick creationists claim that an evolutionary process that occurs by random chance cannot explain how evolution proceeds.

Scientists Respond to the Claim That Random Chance Cannot Explain It All

Again, the problem is an incomplete understanding of what biologists mean. Random chance enters into the evolutionary process in the form of naturally occurring random mutations. These provide the essential raw material for the evolutionary process: genetic variations. Mutations occurring over a 3.8-billion-year period had the potential to create an overwhelming variety of life-forms. The full potential of the mutated variants was not realized in nature, however, for there are many factors that limit the perpetuation of variety.

For example, naturally occurring geologic changes, such as ice-sheet growth, produce changes in climate that can cause extinctions of entire species if the population lacks variants capable of reproducing under the changed condition. In this process, known as natural selection, organisms that survive the limiting factors reproduce successfully, thereby passing on their genetic material to a next generation. In this manner, certain variations that arise randomly in each generation become predominant.

Random mutation is inevitable because mutation is a natural phenomenon. Natural selection is also inevitable because almost any natural population of organisms produces more offspring than can be supported by the limited supply of natural resources.

The "It's Only a Theory" Claim

Evolution is only a theory; it has not been proven.

Science Responds to the Claim That It's Only a Theory

It is undoubtedly true that the ***theory of evolution*** has not been proven. Neither has the ***big bang theory*** nor the ***plate tectonics theory.*** Neither has the ***second law of thermodynamics*** nor the ***universal law of gravitation.*** No scientific theory can ever be proven correct because *all scientific theories are provisional by their very nature.* Scientists never claim infallibility. As Albert Einstein said, "No amount of experimentation can ever prove me right; a single experiment can prove me wrong."

Abundant experimental evidence supports the ***theory of evolution.*** To challenge this theory, it must be shown that this evidence is either incorrect or irrelevant, or that experimental evidence doesn't match its predictions.

The "Planted Evidence" Claim

"Scientific" quick creationists claim that evidence that appears to be contrary to their theory is explained by deliberate creation of that evidence by God. For example, the light that appears to have been traveling from quasars for billions of years was created by God just 6,000 to 10,000 years ago at a point where it would reach Earth in 6,000 to 10,000 years. God also created the world complete with fossils that make it look much older than it really is.

Science Responds to the "Planted Evidence" Claim

For an explanation to be scientific, there has to be a conceivable way to test it. This "planting of evidence" explanation cannot be proven false by any conceivable test; it is unfalsifiable. As such, it does not refute evidence of a multibillion-year-old universe. It is an ad hoc hypothesis in that it introduces a level of explanation more complex than observations warrant.

While this more complex "planting of evidence" explanation might be correct, belief in it remains an article of faith until supporting evidence is found. Since scientists have no way of disproving the more complex explanation, the burden of proof for this idea rests on the "scientific" quick creationists. Similarly, the burden of proof rests on them regarding claims that God created the universe by "special" processes that no longer operate in the natural world, and that the laws of nature by which God created the world are different from those we currently observe.

"Scientific" Quick Creationism: Science or Dogma?

While evolutionary theory has a lot of experimental support, biologists differ as to the mechanisms and pathways by which it occurred. Although significant understanding of these followed the discovery in 1953 of the structure of deoxyribonucleic acid (DNA), the molecule that carries hereditary information for all known organisms, many important aspects of these processes (sources of mutations, importance of various selection processes, and "tempo" or relative rates of mutation and speciation) remain to be fully explained.

People familiar with the nature of scientific thought will realize that debate about these mechanisms and pathways is not a sign of weakness of the ***theory of evolution.*** It is instead a sign of strength in the scientific endeavor. By contrast, unwavering, unquestioning adherence to ***"scientific" quick creationism*** is a telltale sign of dogma.

Evolution and Faith

The Bible is a book of religion, not a book of science. It is therefore inappropriate to hold the Bible to a high scientific standard. It is also inappropriate to read it as a literal account of creation.

When considering matters of science and matters of faith, it is important to keep in mind the demarcation between these two endeavors. Whereas science consists of ideas whose validity is generally supported by experimental evidence, faith consists of beliefs whose validity is not demonstrable by experiments (evidence is irrelevant). Although the two may have common ideas about natural phenomena, faith alone endeavors to transcend those ideas. In this sense, there need not be a conflict between science and religion. Science as well as religion can be a profound source of spirituality. In fact, many people feel that a deeper understanding of nature's wonders actually enriches their faith.

A Cautionary Tale

The most famous case of religious fundamentalists attempting to replace the teaching of the ***theory of evolution*** with a Book of Genesis–derived belief involved a young science teacher named John Scopes and his response to the passage of the Butler Act. This act was adopted by the Tennessee state legislature in March 1925. It made it unlawful "for any teacher in any of the universities, Normals (teacher training schools), and all other public schools of the state that are supported in whole or in part by the public school funds of the state, to teach any theory that denies the story of the Divine Creation of man as taught in the Bible, and to teach instead that man has descended from a lower order of animals."

Passage of this law was part of a fundamentalist movement sweeping across the country. Scopes admitted to violating the law and was arrested and brought to trial. Despite the efforts of his lawyer, Clarence Darrow, who argued passionately and persuasively against opposing lawyer William Jennings Bryan, Scopes was declared guilty and fined $100. (The conviction was later overturned on a technicality.) The *Scopes* decision was upheld on appeal and never made it to the U.S. Supreme Court.

Publishers were so fearful of raising fundamentalist ire that the ***theory of evolution*** disappeared from most U.S. textbooks for the next 35 years. It returned to the public schools upon the alarm raised by the construction and successful orbiting of the Sputnik satellite and the ensuing race to catch up with the Soviets' scientific superiority. The Butler Act was repealed in 1967.

"Scientific" quick creationism was developed in response to these events. Believers attempted to use their pseudohypothesis to show that science supports creation more than evolution. Although their efforts have failed in the courts, they did pay off in the late 1990s in the form of legislation passed in the state of Kansas. The State Board of Education in Kansas voted to omit evolution from science standards for Kansas students. Those supporting the teaching of evolution said in response that the new science standards would hurt students when they pursue higher education, make Kansas a laughingstock, and allow creationism back into public schools. Fortunately, the vote was subsequently overturned and most of the board members who had voted in favor of the omission were defeated in the next election.

Public school teachers must never be subjected to government-imposed requirements that they base their understanding of phenomena on the tenets of a religion. Not only do such requirements violate the separation of church and state, they also seriously retard scientific development essential to maintaining a country's technological and economic strength.

Lysenkoism

A comparable situation occurred in the former Soviet Union. In this case, pursuit of an understanding of phenomena was retarded not by the imposition of a religious belief, but rather by the tenets of a prevailing political ideology. Genetic research was impeded by government-imposed belief in a theory first articulated by J. B. Lamarck, an eighteenth-century French scientist. Lamarck's theory of evolution predates that of Darwin. It states that evolution occurs because organisms can inherit traits that were acquired by their ancestors during their ancestors' lifetimes. This idea was picked up by I. V. Michurin, who passed it on to T. D. Lysenko, a practical-minded agricultural specialist in the Soviet Union, who believed that he had developed improved methods for seed germination and crop production.

While genetic research elsewhere was based on the tenets of Mendelian genetics (traits are acquired and fixed at birth; they are not acquired during one's lifetime), Lysenko insisted on adhering to Lamarck's theory. In support of this position, he argued that Lamarckian genetics was more consistent with Marxism than Mendelian genetics.

Mendelian thought was denounced as "reactionary and decadent." Those who disagreed were declared by the government to be "enemies of the Soviet people." Scientists either succumbed to the wisdom of the party, or were dismissed.

Sustained belief in this ideology resulted in a steady deterioration of Soviet scientific thought and practice. After 1948, it was illegal to teach or do research in Mendelian genetics. High school textbooks contained no information about the role of the cell nucleus and chromosomes in heredity. It was not until 1964 that Lysenko lost his influence on Soviet biology, and this unfortunate chapter in the history of science (and the Soviet Union) came to an end.

Lysenkoism may be dead, but its spirit lives on among the creationists, who advocate government-imposed equal time for creationism and evolution in biology curricula.

DO YOU BELIEVE IN E.S.P.?
NO.
S. Harris

8 Normal Sensory Perception, Extrasensory Perception, and Psychokinesis

Equipped with his five senses, man explores the universe around him and calls the adventure science.
Edwin Powell Hubble

"You're fine, how am I?"
What one psychic said to the other

When people are asked to name the senses by which we perceive the world, the usual response is vision, hearing, smell, taste, and touch. There are actually a few additional normal senses that must be added to complete the list. These are not "extra" senses. They are simply additional ones.

We will examine the mechanisms that science has discovered by which these normal senses operate. Then we will examine purported evidence for some "extra" senses. In the process, we'll uncover flaws that show that these senses are pseudosenses.

Normal Sensory Perception (NSP)

What exactly is a sense? A sense is a physical system that includes a sensory receptor for receiving a particular type of physical or chemical stimulation, and a transducer for translating the stimulation into an electrochemical message that it transmits to the final element in the system, the brain, which receives, organizes, and interprets the message. Ultimately, it is our brain that receives information about the realities of the world. This means that the sensation of smell occurs in the brain, not the nose; the sensation of sight occurs in the brain, not the eyes; and so on. Sensory information can also originate within the human brain.

Detection of Sensory Stimuli

There is a certain minimum amount of stimulation required for detection by the human senses. This minimum intensity of physical energy required to produce any sensation at all in a person is called the absolute threshold. In theory, you would never sense any stimulus below the absolute-threshold level, and you would always sense it above that level. In reality, however, we do not detect particular sensory stimuli at exactly the same levels all the time. One reason for this is that expectations regarding a sensation can affect the likelihood that you will detect it. For example, you are more likely to notice someone approaching the front door when you're expecting a pizza delivery.

Receptor Cells and Transduction

Each sense organ has specialized receptor cells to detect the appropriate type of physical energy or stimulation. The visual system has receptors sensitive to electromagnetic radiation (light waves in the visible part of the spectrum). Your taste and smell systems have receptors for specific molecules from foods or other sources. Other senses, such as hearing, touch, and balance, have specialized receptors to detect mechanical energy from the air, from other objects, and even from within the body.

All sensory receptors transduce (convert) the incoming form of energy (electromagnetic, chemical, mechanical, etc.) into the electrochemical form of energy used by the nervous system. Sensory neurons then carry those electrochemical messages to various parts of the brain for information processing. Visual sensors send impulses to the back tip of the occipital lobe, sound sensors send their messages to another area of the brain located on the top inner fold of the temporal lobe, and so on. Each sensory area of the cerebral cortex (the brain's crinkled top layer) normally "knows how" to convert the electrochemical impulses into the "right" experiences.

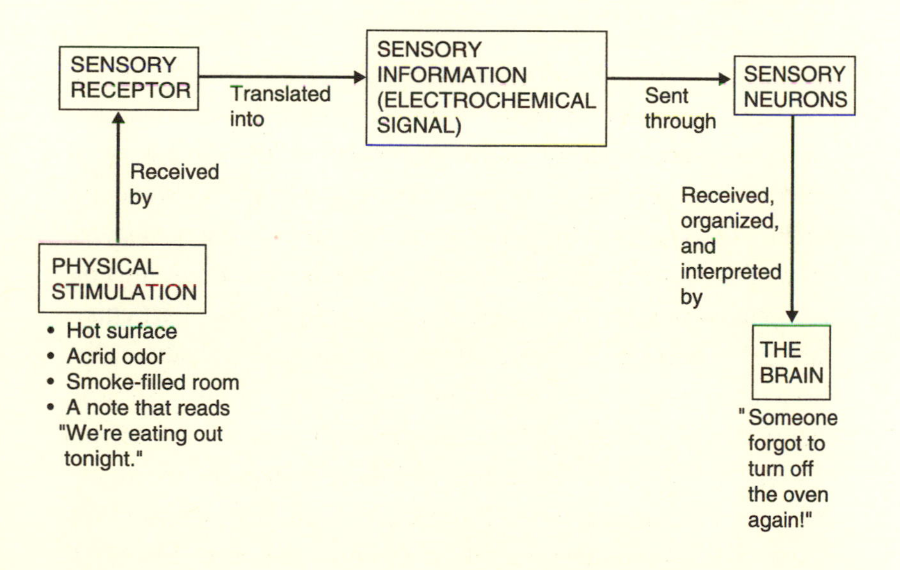

Vision

Light beams that enter the eye reach the retina, a network of neurons on the back surface of the eye. The light beams that enter the eye are electromagnetic waves of different energies, all of which travel at a speed of about 186,000 miles per second. Photoreceptors in the retina (rod and cone cells) transduce the electromagnetic light energy into electrochemical energy that is transported through the optic nerve (sensory neurons) to the brain. The waves themselves are colorless. It is the brain that "interprets" the impulses as "colors." Color is an experience within the mind. It is the experiential finale to a process of reception, transduction, transmission, and interpretation.

Hearing

Sound waves are caused by periodic disturbances, which exert a mechanical pressure or push on air molecules. These air molecules crash into other air molecules, which then crash into still other air molecules, creating a three-dimensional wave of mechanical energy. This wave is transmitted through various parts of the human ear, and eventually reaches the thousands of hair cells located in the cochlea, deep inside the ear. Particular hair cells (hearing receptors) then resonate and transduce the mechanical energy of the sound waves into electrochemical energy that is transported through sensory neurons to the brain.

When a tree in the forest falls and thereby disturbs air molecules, it creates sound waves. If there aren't any hearing receptors nearby to transduce the mechanical energy of the waves produced into electrochemical impulses that register within a human brain, the result is an unheard sound.

Taste

The objects we see are located some distance from our eyes. The events we hear originate some distance from our ears. The items we taste, however, must come in direct contact with us.

To be tasted, a stimulus must contain molecules or charged atoms or groups of atoms that can dissolve in saliva, and there must be enough saliva in our mouths to dissolve these chemicals. The tasty substance lands on the surface of the tongue, which contains clusters of taste receptor cells or taste buds located on small visible bumps. Molecules or charged atoms or groups of atoms of the tasty substance mixed in saliva fit into appropriately sized and shaped depressions within the receptors. Receptors transduce the chemical stimulation into electrochemical energy that is transported to the brain.

There is no "taste" to the substances themselves. They simply activate a process that is interpreted by the brain as sweet, sour, salty, or bitter, depending on which receptors are activated.

Smell

In a process similar to that of taste, when we smell something, we do so by making direct contact with it. Like taste, smell is chemically activated. Scent-bearing molecules in the air are carried into the nasal cavity through either the nostrils or the mouth. They reach small hair cells located high in the nasal cavity. Gas molecules fit into openings in the receptor cells and are transduced into electrochemical impulses that are transported to the brain.

There is no "smell" to the scent-bearing molecules. They simply activate "odor messages" that are interpreted by the brain as smell sensations such as acrid, fruity, and salty, depending again on which receptors have molecules trapped in their openings.

Touch

The sense of touch is another direct-contact sense. Our many-layered skin contains various kinds of sensory receptors, which allow us to detect a variety of sensations associated with the sense of touch. The sensation of pressure results from changes in the skin's shape when objects are pressed against it. The sensation of warmth or cold is a response to the molecular activity of whatever touches our skin.

Too much stimulation of the skin (or other senses) generally causes pain sensations. The pain, however, is not located in the object (a red-hot coal) that causes the pain. The object simply activates a process that is interpreted as pain. Pain can also be created by stimulation from inside our bodies, for example, damage to internal tissues located where there are pain receptors can result in headaches or back pain even though the pain receptors are not located in the back or head.

Position

A sensory capacity that we usually take for granted is our ability to know how and where our bodies are positioned in space. This ability includes awareness about where various parts of our body are in relation to one another, and also how our bodies are positioned in regard to the pull of gravity. The body needs these senses to make almost any intentional movement.

The kinesthetic sense helps a person be aware of skeletal muscle movements. This sense operates through kinesthetic receptor cells located primarily in our joints, but some kinesthetic information also comes from muscles and tendons. These receptors detect changes in the movement or position of our muscles and joints.

They transduce this mechanical energy into electrochemical energy that travels through pathways in the spinal cord and eventually reaches the brain. We become aware of the existence of this sense only when it is absent, for example, when our leg "falls asleep" and we have trouble walking.

The other position sense is the vestibular sense, which tells us about balance, about where we are in relation to gravity, and about acceleration or deceleration. This sense is determined by the position and movement of the head, relative to a source of gravity. We detect vestibular sensations through hair cell receptors deep inside our inner ears. When stimulated, these receptors send neural impulses to the brain. When overstimulated, they can produce feelings of dizziness and nausea that are aptly referred to as motion sickness.

The Senses: Windows onto the World

Our knowledge of the real world is limited by the limitations of our senses. Not only are our natural senses limited by their need for a minimum amount of a sensory stimulus before being able to detect it, they are also limited in the range of signals they can detect.

The visual "window" open to us in the electromagnetic spectrum is limited to wavelengths from about 400 to 700 nanometers. These wavelengths correspond to colors in the visible spectrum ranging from violet to blue to green to yellow to orange, and then red. This range can be extended with the use of special devices such as night-vision goggles that allow us to detect wavelengths in the infrared region (about 20,000 to 60,000 nanometers). Some animals possess a wider range of vision than humans. Snakes have sensors in organs lining their lips that let them see heat patterns made by mammals.

The audio "window" open to normal hearing is limited to a frequency range between 20 and 20,000 cycles per second. Signals outside this range can be detected with special devices such as those used in the field of medicine, where ultrasonic sound waves (above 20,000 cycles per second) are used for diagnosis. Reflection of these waves from regions in the interior of the body can be used to detect a wide variety of anomalous conditions such as tumors, and to study various phenomena such as heart-valve action.

Our visual window opens onto only a small portion of the electromagnetic spectrum. Our audio window opens onto only a small portion of the sonic spectrum. Likewise, our chemical window opens onto only a small portion of the vast array of molecules that reach our tongue and nasal passages. It is understandable that people try to devise ways to widen their visual, audio, chemical, and other windows onto the world, and in addition, wonder whether there are unknown windows that might reveal other aspects of the world.

There are people who claim additional windows do exist that make possible a number of "extra" senses. Let's examine those claims.

Psychology of Paranormal Phenomena (Parapsychology)

Parapsychologists who study these alleged phenomena use the word "Psi" to denote what they call extrasensory perception (ESP—perception not using the normal senses of sight, hearing, taste, smell, touch, and position) and psychokinesis (PK, also called telekinesis—the production of motion in physical objects by the exercise of mental powers).

The word *psychic* has become a generic term referring to the particular abilities and attributes of persons claiming to be able to manifest psi. Psychics are people said to possess powers of cognitive insight (ESP) and physical manifestation (PK).

How one acquires these powers in the first place is said to vary from individual to individual. Some people claim they were born with them. Some attribute them to a traumatic experience or an accident. Others attempt to attain psi through psychic training seminars and courses. The demand for such programs is so great that psychics have become big business.

Even the military has been interested in attaining psi. In the 1960s, the Pentagon spent millions of dollars for psychic research in the hope that it could unleash the military potential of psi. It knew the Soviets had conducted psychic research aimed at the deployment of psychic weapons. America was eager to close what it perceived as an ESP gap.

ESP: Distance Learning

Many people claim to have had psychic experiences. They thought of a friend just moments before the friend telephoned; they had a premonition a plane would crash, avoided the next flight, and later learned the plane had crashed; they dreamt of winning the lottery and then won it. Though true, these experiences prove nothing about psychic ability. They are merely odd coincidences that command our attention. What we ignore are the far more frequent times when we think of a friend, but don't hear from her, when we believe a plane will crash and it doesn't, and when we dream of winning the lottery (again) and all we have to show for it is a pleasant dream (again).

There are a variety of these alleged extrasensory perceptions:

TELEPATHY Psychic knowledge of someone else's thoughts or feelings

CLAIRVOYANCE Psychic awareness of an unknown object or event

PRECOGNITION Psychic knowing of future events

RETROCOGNITION Psychic knowing of past events

PSYCHOMETRY Psychic ability to learn the history of an object

Anecdotes are insufficient as scientific evidence of these perceptions. What is needed are controlled experimental tests that rule out the possibility of coincidence. Classic experiments involving such phenomena were carried out at Duke University beginning in 1929 by Dr. Joseph B. Rhine and his wife and collaborator Louisa. The Rhines used a 153 of cards designed by their colleague Carl Zener. Each card contains one of five geometric symbols: a cross, a star, a circle, wavy lines, and a square. Five cards of each symbol are combined to make the 25-card Zener deck. Rhine tried to determine whether it was possible for a subject to correctly identify the symbols on the cards without having any sensory contact with them. Rhine, by the way, is the person who coined the terms *extrasensory perception* and *parapsychology.*

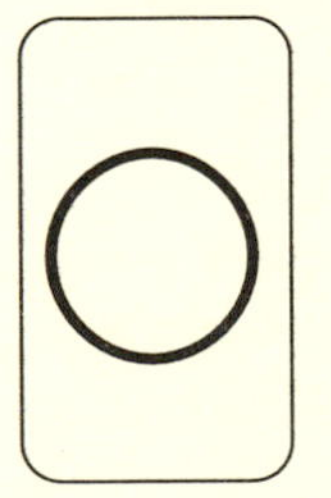
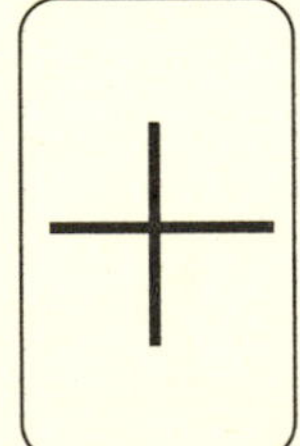
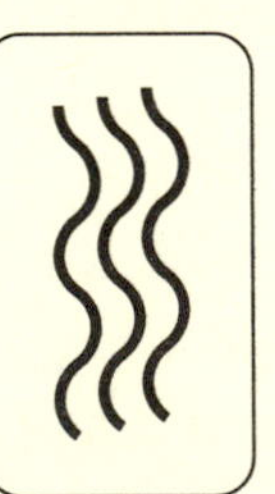
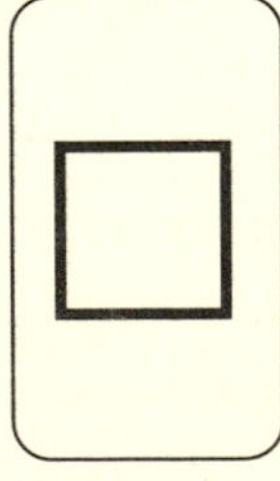

Here are descriptions of some of the card-guessing experiments conducted using this deck. The first three test clairvoyance.

THE SINGLE CARD CALLING TEST The symbol on the top card is guessed, removed face down, followed by the next card, and so on throughout the deck.

THE BLIND MATCHING TEST One card of each symbol is placed face down. The position of the cards is unknown. The five cards are shuffled and the subject is asked to guess the order of the symbols.

THE PACK CALLING TEST The subject makes 25 consecutive calls directed at a shuffled but unbroken deck located in another room.

The next two test telepathy.

THE GENERAL TELEPATHY TEST The sender shuffles the cards, cuts them, and looks at the face of each card, while the receiver attempts to read the mind of the sender and guess the symbol on the card on which the sender is concentrating.

THE PURE TELEPATHY TEST The sender chooses a random order of cards and memorizes them. The receiver then attempts to guess the symbols.

The sixth one tests precognition.

THE PRECOGNITION TEST The subject writes down ahead of time the order the cards will be in after having been shuffled and then guesses the order after they have actually been shuffled.

Since one out of every five cards contains a particular symbol, the chance of correctly guessing a card is 1 in 5 (or 5 out of the 25 cards in an entire deck). This translates to a probability of 20 percent or 0.2. Subjects that consistently do better than 0.2 are considered to possess ESP ability.

Rhine reported that many subjects had scored above 0.2, with the most successful one scoring 0.32 over 17,250 trials. The odds against these results being due to chance are so great as to nearly rule out random chance as an explanation for them. By 1934, Rhine was convinced he had overwhelming evidence of ESP. A number of other psychology departments repeated his experiments in an attempt to confirm the results, but none were successful.

In 1940, Rhine coauthored a book, *Extrasensory Perception After Sixty Years*, in which he suggested that something more than mere guesswork was involved in his experiments. He was right! It is now known that the experiments conducted in his laboratory contained serious methodological flaws. Tests often took place with minimal or no screening between the subject and the person administering the test. Subjects could see the backs of cards that were later discovered to be so cheaply printed that a faint outline of the symbol could be seen. Furthermore, in face-to-face tests, subjects could see card faces reflected in the tester's eyeglasses or cornea. They were even able to (consciously or unconsciously) pick up clues from the tester's facial expression and voice inflection. In addition, an observant subject could identify the cards by certain irregularities like warped edges, spots on the backs, or design imperfections.

Sometimes, results were falsified. One example of falsification is that perpetrated by Walter J. Levy, the director of Rhine's Institute of Parapsychology, who was discovered to be producing significant-looking results through the manipulation of data-recording equipment. Another is that of S. G. Soal, who claimed to have replicated

Rhine's experiments and results in his own laboratory. One of the people who had assisted with Soal's experiments, Gretl Albert, later stated that she had observed Soal altering numbers on the record sheets. The targets were the digits from 1 to 5. She specifically indicated that she had seen him changing the numeral 1 on the target list into numerals 4 and 5 during studies done on a man named Basil Shackleton, who had seemed to demonstrate in thousands of tests that he had genuine ESP powers. Guided by this allegation, subsequent analyses of the full record confirmed that there was an excess of hits when the target was a 4 or a 5, and a shortage of ones on those trials in which the guess was a 4 or 5. In the 1970s, Zener cards were largely replaced by testing techniques using random-number generators, and others using more complex and meaningful targets such as paintings and photographs.

Reports of badly flawed experiments that are wide open to cheating are not uncommon. Here's another example. When Israeli psychic Uri Geller was tested under conditions set up by Targ and Puthoff, he was required to make a drawing corresponding with a target randomly taken from a dictionary (a test of "remote viewing"). Geller was able to identify 7 of the 13 targets, a remarkable 54 percent success rate. When Targ and Puthoff suggested that something other than mere guesswork was involved in Geller's feat, they too were right. In independent tests under rigorously controlled conditions set up by Rebert and Otis, Geller failed to identify one target in the whole series.

To avoid the methodological criticisms that have plagued psychical research, the following precautions have been recommended.

- To eliminate sensory cues, targets (hidden materials the subjects are attempting to identify) are handled as little as possible so that random scratches or markings do not become the basis for the subject's responses.

- Targets are prepared by an independent assistant who has no contact with the subject (the "double-blind" method).
- Random selection and presentation of targets must be ensured through the use of random-number tables or other random sources as the basis of target sequences.
- Appropriate procedures must be designed and followed to ensure that the subject has no opportunity to cheat. The subject can never be left alone with the targets in a clairvoyance experiment, and cannot be allowed to communicate with the receiver in telepathy experiments. Targets must be concealed from the subject by full-length screens or opaque envelopes, or kept in a place to which the subject has no access.
- The experimenter who interacts with the subjects must not know what the targets are on any given trial.
- Scoring is double-checked by an assistant who does not have information about the hypothesis of the experiment, had no contact with the subjects, and does not know to which experimental group or condition the subjects belong.
- The statistics used to evaluate the data should be evaluated independently by statisticians to ensure their appropriateness.

Dowsing

The alleged psychic ability of dowsing is said to enable dowsers to locate underground substances or objects. These include underground water, minerals such as oil, treasure, archaeological remains, and even dead bodies! Dowsers use a forked piece of hazel, rowan, or willow wood, a Y-shaped metal rod, or a pendulum or object suspended by a nylon or silk thread, in their attempts to locate buried materials.

The dowser holds the two branches of a dowsing rod (one in each hand) and points the device skyward and away from the body. As the dowser walks near a promising site, the rod bends downward until it points to the site or quivers violently. Pendulum dowsers usually hold the pendulum at arm's length and note that the pendulum swings back and forth when the dowser is over the substance being searched for. Map dowsers claim to be able to locate substances by moving their pendulum over the surface of a map. Dowsers believe they receive transmissions from the hidden object that cause involuntary muscle contractions, which in turn make the rod or pendulum move.

Do dowsers occasionally locate hidden substances? Yes. Do these events support the hypothesis that dowsers can detect hidden substances better than through chance guessing and without the use of clues in the environment? No.

Controlled experiments set up to test the abilities of dowsers have shown that dowsers are no better at finding hidden substances than chance would predict. Knowledgeable scientists (and knowledgeable dowsers) can use surface clues such as surface water, vegetation, and soil color to locate likely sites. Furthermore, if a well is drilled in an area where water is likely to be found, it will likely be found!

What does cause the muscle contractions that move the dowser's rod or pendulum? The movement is caused by suggestion and unconscious muscular activity in the dowser. It has been demonstrated that just thinking about a certain physical action (like a dowsing rod tilting downward or a pendulum swinging) causes minute reactions in the muscles that would be used in such actions. And, the slightest movement in wrists or hands will be magnified in the movement of a rod or pendulum. As with Ouija board operators, the individual has no conscious awareness of this phenomenon and may be genuinely surprised by it.

Nostradamus

Throughout history, people have consulted a variety of seers in an effort to be forewarned of events to come. Insights gained could presumably be used to avoid mishaps and tragedies. Heads of state could avoid sudden takeovers or assassinations. Even today, people act upon advice they purchase in person, by mail, or over the telephone from psychic advisers (also called clairvoyants) claiming this ability. Use of this information is somehow supposed to be able to "change the future" (whatever that means).

One of these psychic advisers has been dead for nearly five centuries! He is Nostradamus (Michel de Notredame) who wrote predictions in the form of almost a thousand verses. His four-line verses or quatrains, each of two rhymed couplets, were written in groups of 100, known as Centuries. *The Centuries*, his collected verse, was probably first published in 1555. Although most of the references in these verses refer to events and places in France in the sixteenth century, after Nostradamus's death, people began to take advantage of his reputation and use the verses to foretell historical events occurring in other parts of the world and in other centuries.

In this manner, Nostradamus, who lived from 1503 to 1566, has been credited with predicting both World Wars, the atomic bomb, the rise and fall of Hitler, the assassinations of the two Kennedy brothers, AIDS, and more. Nostradamus himself said they were deliberately puzzling and cloudy. As a result, their wording leaves them wide open to various interpretations. Furthermore, Nostradamus's works have been subject to counterfeiting and alteration to suit the purposes of the church, governments, and others who would interpret them to suit their own ends. For example, during World War II, an astrologer named Louis de Wohl was employed by the British to compose 50 bogus "Nostradamus" quatrains predicting Germany's defeat. Written in German, these were smuggled into Germany in the form of astrological leaflets.

Why Psychic Readings Only *Seem* to Work

Psychic readings providing information about the past (retrocognition) as well as the future (precognition) can seem to be surprisingly accurate. The reasons for this become clear once the techniques of social and psychological manipulation used by clairvoyants are understood. Readings done in person make it possible for the reader to gather background information from the client's general appearance (clothing, apparent state of health, etc.). Important clues can be obtained from birthstone rings or zodiac jewelry that disclose the client's approximate birthdate. The state of hands and fingernails can provide information about one's occupation or hobbies. A pale band on the finger where a wedding band was once worn not only speaks of the past, but points to the future. Forgotten scars speak of minor or major accidents. Then too, opening remarks and subsequent conversation allow the reader to judge the client's level of education and mood.

Although we tend to think that our own life situation and problems are unique, we all go through more-or-less similar stages in life and encounter concomitant problems associated with them. The fortune-teller can utilize the initial observations, take into consideration the client's age and sex, and make a reasonable guess about that person's past experiences and present problems. Subsequent vague statements and leading questions can glean additional information and either confirm or eliminate suspicions.

With these insights in mind, the fortune-teller can create an analysis likely to hit the mark (pun intended). The assessment is couched in general and ambiguous terms that can apply to almost anyone (much like that of an astrological horoscope). "Your aspirations are sometimes unrealistic." "At times you are extroverted, at other times you are introverted." "Sometimes you doubt whether you have made the right decision." "You have a great deal of unused capacity." "You have a strong need for other people to like you." Who could disagree?

The client has come to the fortune-teller because she needs advice. She has a need to believe the reading. Under such conditions, there is a willingness to try to make the general and ambiguous disclosures fit her own situation. If, however, the feedback is not accepted by the client, these fortune-tellers have a ready excuse. They inform her at the outset that success depends on her cooperation. If the feedback doesn't succeed, they tell her that she didn't cooperate!

Psychokinesis: Action-at-a-Distance

You pick up the remote control device and with the press of a button turn on a television set located on the other side of the room. Since there is no physical connection between the device and the TV, what you have done is cause an action-at-a-distance to take place. What makes this action possible is the generation of an electromagnetic signal that travels through the air from the remote control to the TV set. Although such events may seem mysterious to the uninitiated, the "invisible" electromagnetic signal can be detected using scientific instruments.

Action-at-a-distance is a well-known and well-studied phenomenon. It can also involve gravitational forces, as when a cannonball released from the top of the Tower of Pisa is attracted to the ground below, and magnetic forces, as when magnets and your refrigerator are attracted to each other.

Parapsychologists claim that action-at-a-distance can be accomplished using only the power of the mind. They claim that objects can be moved and even have their shapes changed by psychic means alone. One of the most famous demonstrators of psychokinesis is Uri Geller. His trademark trick is bending keys and cutlery "using only the power of his mind." This feat has been observed on television by millions of people worldwide. And it certainly has seemed to many of those viewers that no physical means had been utilized.

In reality, however, Geller, an experienced magician and showman, simply bends the objects when no one is watching. But, you may argue, millions of people were watching him on TV! Geller is a master at an essential tool of the magician: misdirection or distracting peoples' attention. He is quite good at projecting an air of innocence that belies his actions. That he can fool so many people is a tribute to sleight-of-hand artistry, not psychic power.

Scientists are rarely trained as magicians and have often been conned by demonstrators of psychic phenomena. Subjects with ulterior motives who are in a position to take advantage of a loose protocol usually succeed. The same Dr. Rhine who studied ESP, also studied and felt he had evidence for PK. Attempts to replicate Rhine's findings under controlled conditions all failed. Successful tests of PK reported by him were the result of inadequate controls or falsification of data.

The case of reputed psychics Steve Shaw and Michael Edwards demonstrates dramatically how easy it is to fool otherwise intelligent investigators. Tests carried out over a period of two years in the early 1980s at the McDonnell Laboratory for Psychical Research at Washington University in St. Louis, Missouri, covered a large range of ESP and PK experiments. In tests designed to be as controlled as possible, they demonstrated the ability to visualize a picture contained in a sealed envelope. In this instance, the controls were so loose that they were able to simply remove the staples sealing the envelope and peek inside.

Shaw and Edwards were actually amateur magicians planted in the laboratory by James Randi, himself a magician, to show that without the correct controls and a strict protocol, it is possible to create the illusion that one possesses psychic powers. The deception continued undetected throughout the two years. It was later revealed that Shaw, Edwards, and Randi had agreed in advance that they would never allow the deception to proceed to a point where McDonnell Laboratory submitted a report about its "findings" for publication.

Up, Up, and Away: Levitation

Levitation is said to result from powers of psychokinesis. Levitation acts—including people rising in the air unassisted, flying through the air horizontally, and climbing a rope into the air until they disappear from view—seem to be in defiance of the ***law of universal gravitation.*** Among the people who have claimed to have levitated are spiritualists, Indian fakirs, and members of the Transcendental Meditation Movement (TMM).

In reality, levitation is a clever stage illusion in which parts of the body are supported by platforms that are not in the line of sight of the audience or by wires that cannot be seen because they are transparent or very thin. Photographs of TMM members hovering above the ground are actually photographs of TMM members in the lotus position bouncing up and down on a mat. The photographs are taken near the top of the bounce.

If Psi Effects Are Ever Demonstrated Beyond a Reasonable Doubt, How Might They Be Explained?

One of the reasons scientists have difficulty believing that psi effects are real is that there is no known mechanism by which they could occur. PK action-at-a-distance would presumably employ an action-at-a-distance force that is as yet unknown to science. Is it possible such a force exists and has not yet been detected? Yes, but until detected, it cannot be used to explain how PK might work. Similarly, there is no known sense (stimulation and receptor) by which thoughts could travel from one person to another or by which the mind could project itself elsewhere in the present, future, or past.

Is It Possible to Prove That Psi Does Not Exist?

The standard of science for acceptance of a claim that an entity or phenomenon exists is that the claim be established beyond a reasonable doubt. This means that claims that psi has been demonstrated must prove reproducible by a broad range of investigators. Claims for the existence of psi have not met this standard.

No amount of evidence (or lack of evidence), however, can ever prove that an entity or phenomenon does not or could not exist. *It is impossible to prove a universal negative.*

Extrasensory perception and psychokinesis fail to fulfill the requirements of the scientific method. They therefore must remain pseudoscientific concepts until methodological flaws in their study are eliminated, and repeatable data supporting their existence are obtained.

"WHAT'S NICE ABOUT WORKING IN THIS PLACE IS WE DON'T HAVE TO FINISH ANY OF OUR EXPERIMENTS."

9 Reflections on the Scientific Approach to Reality

Man prefers to believe what he prefers to be true.

Francis Bacon

Fool's Gold or Real Gold?

People who see shiny golden-colored flakes in a rock they just picked up are sometimes led to believe they have struck paydirt. That's because genuine gold and minerals such as pyrite, the most famous "fool's gold," are both yellow, opaque, and have a metallic luster. Whereas genuine gold is relatively rare, pyrite is so common in Earth's crust that it is found in almost every environment, and hence has a vast number of forms and variations. Its golden look, beautiful luster, and interesting crystals have made it a favorite among rock collectors.

Although these two minerals are deceptively similar, there are telltale clues to their identity. One of the simplest tests is called a "streak test," in which a mineral sample is rubbed against a piece of white unglazed porcelain. Gold is soft enough to leave a golden streak on the plate. An imposter such as pyrite, which is actually a compound of iron and sulfur, leaves a black streak. *All that glitters is not gold!*

Pseudoscience or Real Science?

Pseudoscience camouflaged as science also appears in many guises: belief in alien spaceships and abductions by alien life-forms, out-of-body experiences and entities, astrology, "scientific" quick creationism, as well as ESP and PK. At the beginning of this book, we provided a "streak test" in the form of telltale clues to help identify and strip away that camouflage. The list of clues consisted of potential flaws in the application of the scientific method. Here are summaries of what those clues have revealed.

Alien Spaceships and Abductions by Alien Life-Forms

Observations of flying objects identifiable as alien spaceships and reports of abductions by alien life-forms are filled with classic signs of pseudoscience, namely, personal anecdotes by untrained observers who exaggerate, mistake, or imagine phenomena. Hypotheses based on such observations are unreliable to begin with, and are far more complex than the observations warrant. Predicted recurrences of the phenomena lead to experiments replete with the flaws inherent in the observations from which the hypothesis arose. Reluctance to recycle ideas about alien life-forms inhibits the search for the truth about these phenomena.

Out-of-Body Experiences and Entities

Out-of-body experiences and entities are observed by people whose imaginations have gotten the best of them, by people in an altered state of consciousness, by people who report the phenomena for ulterior motives, and by people who have been deliberately deceived by con artists. Hypotheses based on such observations are unreliable to begin with, and are more complex than observations warrant. Experiments that test predictions based on them are fraught with the observational problems that gave rise to the hypothesis in the first place. Again, recycling is inhibited by wishful thinking that these phenomena are real.

Astrology

The original observations that led to astrological beliefs are so out-of-date and inaccurate that the hypothesis suffers from the same limitations. This vague and inappropriately general hypothesis allows such a wide margin of error that its predictions cannot be evaluated definitively. Since the hypothesis helps fulfill people's quest for easy answers, it is difficult to displace.

"Scientific" Quick Creationism

Observations on which "scientific" quick creationism is based come directly from the book of Genesis, a book of religion and not of science. It is therefore inappropriate to base a scientific hypothesis upon them. Subsequent arguments on behalf of this hypothesis face a steep uphill battle because they run counter to an interlocking network of well-established scientific hypotheses.

ESP and PK

Our understanding of *normal* sensory perception is based upon observations that are reproducible. Hypotheses about the nature of these phenomena can readily be put to the test by examining the results of experiments based on their predictions. On the other hand, observation of *extra*sensory perception and psychokinesis is tenuous at best. These hypotheses are difficult to evaluate because of the questionable nature of the phenomena they purport to explain. Wishful thinking that people possess "extra" powers inhibits the search for the truth about these alleged phenomena.

Bigfoot and Nessie: Observations or Pseudo-Observations?

Scientists must always be open to new information and ideas. For example, they are continually on the lookout for previously unknown animals. Although most of the previously unknown animals are insects and small animals, in the past decade, scientists have turned up a deer species, wild ox, 10 new species of primates (including marmosets, tamarins, and a capuchin monkey), as well as an antelope.

Reports of sightings of unknown animals must be carefully evaluated before the existence of such animals is confirmed. Here are two accounts of scientists' attempts to determine whether reported creatures are realities or illusions.

Big "Foot" Prints

Bigfoot, also called Sasquatch, is variably described as a 6- to 15-foot-tall humanlike creature with brownish-red (sometimes tan or black) fur that walks upright on two feet, often giving off a foul smell, and either moving silently or emitting a high-pitched cry. The large and deep footprints attributed to this heavy creature have measured up to 24 inches in length and 8 inches in width. It seems to represent the North American counterpart of Asia's Yeti, or Abominable Snowman.

Hoaxes contribute to many of the sightings. In 1976, four youths admitted to having taken turns dressing up to resemble Bigfoot and making Bigfoot "tracks" in Wisconsin, using wooden attachments on their shoes. A pair of boots found in Arkansas in the late 1970s had pieces of rubber tire cut in the shape of large feet, attached to the soles. In 1982, Rant Mulleno in the Pacific Northwest admitted that he had been making hoax Bigfoot footprints for 50 years, using Bigfoot "feet" carved from wood.

Although a number of sightings (including photographs and films) are definitely hoaxes, many sightings are probably not hoaxes. For example, a person sees "something" in the woods, doesn't really get a good look at it, has heard about Bigfoot, and interprets the "something" as the genuine article.

In the final analysis, it will never be possible to prove that Bigfoot does not exist. If it does exist, however, how could such a large and strange-looking creature remain so well hidden for long, and why has something more tangible in the way of evidence, such as bones or a skull, never been found?

Loch Ness Monster

Millions of people have traveled to Loch Ness, an extremely large, deep, and cold freshwater lake located in northern central Scotland, in hopes of sighting the Loch Ness Monster, or "Nessie," as she is affectionately known. Many return home convinced that their hopes have been fulfilled.

Nessie has been described as a dinosaurlike beast whose long neck and small head emerge from the murky waters of the lake. A number of sightings are accompanied by photographs. These are always very gray and grainy, with many shadows and outlines. In some, what appears to be the back of the creature can be seen breaking the water. Although many of the photographs are faked, some are genuine. Are they evidence of Nessie or simply photographs of logs, shadows on a wave, driftwood, or even groups of seals traveling in single file?

To date, no physical remains or other traces of Nessie have been found. Five separate investigations using sophisticated sonar equipment to track her produced no evidence to support Nessie's existence. What is certain is that the Loch Ness area is the site of a lucrative tourist industry, complete with submarine rides and a multimedia tourist center.

Spontaneous Human Combustion: Phenomenon or Pseudophenomenon?

Reported phenomena must also be carefully evaluated to establish whether they represent reality or illusion. Spontaneous human combustion, the supposed process in which a human body suddenly bursts into flames as a result of heat generated by internal chemical action, is one such phenomenon.

Reports of fire originating *within* the human body have never been validated. When sudden human combustion does occur, it is always the result of fire from without. This would be the case when someone wearing flammable nightclothes and in an alcohol stupor or under the influence of sleeping tablets falls asleep in an overstuffed armchair and accidentally drops a lighted cigarette on the chair. Another source of ignition is a person who murders and intentionally torches the corpse. Yet another possibility is that of elderly persons igniting themselves accidentally.

Conditions within the human body are simply incompatible with internal combustion. Human bodies are 60 to 70 percent water, which is noncombustible (as demonstrated by entertainers who swallow fire with no ill effects). Ingested alcohol is combustible but a person would die of alcohol poisoning long before consuming enough liquor to have even a slight effect on the body's combustibility. The only two combustible substances inside the human body are fat tissue and methane gas. Even if a mechanism was available for ignition of internal methane gas, there's not enough of it to bring human fat to its ignition point. And anyway, all that water would just extinguish the fire.

That burning sensation in your throat is just stomach acid in your trachea.

THE RESULTS OF...

SPONTANEOUS COMBUSTION AND SPONTANEOUS LIQUEFACTION

Explanations of Fire Walking: Hypotheses or Pseudohypotheses?

Once it has been established that a reported phenomenon is real, proposed hypotheses about that phenomenon must be evaluated to determine whether any are pseudohypotheses. Consider the case of fire walking, a real phenomenon in which people walk barefoot across red hot embers and emerge unharmed.

It takes several hours to prepare a 10-foot by 30-foot fire-walking trough. Large quantities of wood must be burned until they are reduced to red hot embers. The temperature in the middle of the path reaches 800°F (paper burns at 451°F; at 800°F, steak quickly becomes "well done"). Nearby spectators perspire profusely because of the intensity of the radiated heat. A man removes his shoes and socks, steps barefoot onto the trough, and shows no sign of pain as he walks briskly across. He emerges without any blisters or burns on the soles of his feet. In countries such as Singapore, Malaysia, Fiji, Sri Lanka, and India, fire walking is part of a religious ritual associated with mystical powers. It has also been touted as a test of positive thinking.

Is this phenomenon a miraculous feat by miraculous feet? Or can it be explained scientifically? After all, if you touch a metal cake pan in an oven where the temperature is only 400°F, the pan will surely burn your skin. Well, fire walking requires neither mystical power nor positive thinking. It is instead, a dramatic demonstration of a natural physical phenomenon.

If you stick your hand inside an oven in which the temperature has reached 400°F and touch the cake pan, you will be burned. But, if you stick your hand into the oven and just "touch" the hot air or touch the cake, both of which have also reached a temperature of 400°F, you will not be burned. The explanation for this is fairly simple. Cake and air are poor retainers of heat (they have a low heat capacity), and they do not conduct heat very quickly (they have poor thermal conductivity). As long as you don't stay in contact with cake or air for too long, not enough heat is transferred to your hand to burn it. Metals (like the pan), on the other hand, are excellent retainers of heat and excellent conductors of heat. As such, they can rapidly transfer sufficient heat to do damage.

The ashes that coat the surface of hot embers, even at 800°F, are poor retainers of heat, and, like cake and air, they do not conduct heat very quickly. Thus they provide a layer of insulation between the solid embers and the soles of the fire walker's feet. The "secret" of successful fire walking is to walk fast enough so that each foot is in contact with the embers for only about a second before being lifted. The entire walk generally lasts less than 7 seconds.

In other words, if you walk too slowly, you will suffer the agony of defeat. Ouch!

Where Reality Ends . . . and Illusion Begins

Beliefs and hopes should be based on critical rather than wishful thinking. Pseudoscientific beliefs impede progress toward such a reality-based view of the natural world because the people who adhere to them do not engage in critical thinking. The road to illusion that leads to these beliefs is therefore a road to delusion: false belief and false hope held in spite of invalidating evidence.

Alternative medicine provides many examples of such roads to delusion. People are tempted to follow them because of significant problems associated with traditional (conventional) medicine. Although traditional medicine has been enormously effective in alleviating the physical ills of humankind, it can involve painful procedures that promise, but do not guarantee, relief from physical problems. It can be expensive. Risk of malpractice, though low, is real. Prescribed pharmaceuticals can have uncomfortable and sometimes unforeseen side effects. Traditional medicine may not even be able to discover the cause of an illness or mitigate the pain associated with that illness.

No wonder people are tempted to reject traditional medicine and seek relief from practitioners of alternative (nonconventional) health care. No wonder people are enticed by methods that promise to be less invasive, less expensive, less risky, and less frightening. But, caveat emptor, let the buyer beware of treatments that currently lack confirming evidence.

Pseudoscientific methods, although they sometimes *seem* to work, do not *really* work. The fact that someone underwent a certain treatment and subsequently felt better does not necessarily mean the treatment caused the improvement: correlation does not imply causation. Treated or untreated, many diseases will simply run their natural course; they are naturally self-limiting. Rare, but not impossible, naturally occurring spontaneous remissions can and do occur, even with diseases that are frequently lethal.

In addition, psychological relief attained through belief in pseudoscientific techniques can be misinterpreted as physiological relief. Furthermore, actual physiological relief can be attained from a new treatment *even if the treatment is inert!* This ability of our bodies to sometimes heal what ails us, if only we believe in the cure, is known as the "placebo effect." For example, asthma sufferers have been told a new inhalant will open their airways, and that's exactly what happens, even though the inhaler contains a "placebo" or inert substance. In fact, virtually anything that sends a patient one of four messages—someone is listening to me; other people care about me; my symptoms are explainable; my symptoms are controllable—can bring measurable improvements in health.

Placebos behave much like "real" drugs. They have affected cholesterol levels and exhibited cumulative and time-dependent effects. They can also elicit adverse reactions in what are termed "nocebo effects." There have even been rare cases of placebo addiction. Of course, these difficulties in evaluating the effectiveness of pseudoscientific techniques also apply to scientific ones. Therefore, *all* techniques must be evaluated in properly controlled clinical trials that have objective measures of success.

Let's take a closer look at a few of these pseudoscientific alternatives.

Psychic Surgery

Surgery has long been recognized as a necessary and effective mode of treatment of injuries, diseases, and other disorders. With increased knowledge of anatomy, anesthesia, and asepsis (freedom from contamination by pathogenic organisms), present-day surgery has improved greatly from the days when patients died from surgical infections or suffered unnecessarily during the procedure.

Nevertheless, because the procedure has its drawbacks, desperate people are tempted to employ the services of so-called psychic surgeons. These pseudoscientists offer surgical procedures that require no anesthetic, no sutures, no lengthy postoperative healing process, no chance of postoperative shock, no need for X rays or CAT scans, no blood transfusions to replace lost blood, no measures to prevent complications such as postoperative lung infections and blood clotting in the legs, and no endoscopes (flexible fiber-optic tubes equipped with a light and a video connection) inserted into bodily passages to provide views of the interior of hollow organs or vessels. *There's only one problem: no cure!*

Psychic surgery is a pseudoscientific procedure in which the practitioner claims to cure organic ailments by parting and reaching through the skin with his bare hands (no scalpels!) to remove tissue often claimed to be tumorous. When his fingers reappear, amidst a blood-like fluid, they are clutching what appears to be the sought-after diseased human tissue. Seemingly miraculously, the "wound" heals instantly, without even leaving a scar. The patient goes home seemingly cured.

What really happens is that the psychic surgeon has hidden in advance a supply of "blood" (chicken, pig, or cow blood, or a dye made from betel nuts) and tissue (usually fat and sinew from a small animal). Using sleight-of-hand, he conceals these materials in the hollow tip of a rubber false finger that fits over his thumb and which can be concealed in gauze bandages used to cleanse the "incision." In another method, assistants slip the material to the surgeon in plastic vials.

The surgeon simulates making an incision by forming a crease in the patient's skin while squirting "blood" along the fold. Fingers that appear to enter the skin are merely pushed into the fold or simply bent under so they appear to be inside the patient's body. The palmed tissue is then made to appear as if it is emerging from the body itself when the fingers are withdrawn from the fold. And, of course, when the area from which tissue seemed to appear is wiped clean by the surgeon, the skin appears to be restored to its original unwounded condition!

Psychic surgery is nothing more than magic, the deliberate use of physical means to create changes or illusions that make it appear as if a magician has supernatural or paranormal powers. It pretends to be magick, a pseudoscience that purports to contravene the laws of natural science and cause changes in accordance with the will by nonphysical means.

Thousands of people fall prey each year to the false promises of psychic surgeons. The most popular locales for psychic surgery are the Philippines and Brazil. Any relief attained by these bogus cures can be attributed to the same factors that apply to other forms of pseudoscientific alternative medicine. The ultimate tragedy is that victims of pseudosurgeons may refuse to see a regular physician until their disease is so advanced that it is no longer operable.

Crystal Healing

The lovely and often soothing external appearance of crystalline solids is a reflection of their regular, repetitive, three-dimensional molecular structure. Pseudoscientists claim that crystals, especially quartz, can act as "healing centers" through wishes and thoughts of good health that are "locked in" the crystal and enhance one's health when worn. They say crystals work by receiving and then "locking in" thought vibrations (thought patterns).

This claim is inconsistent with what scientists know about quartz crystals and about brain waves. While thin slices of quartz vibrate (resonate) at extremely fast frequencies (millions of cycles per second), brain-wave patterns have markedly different frequencies of only eight to several hundred cycles per second and therefore are incapable of inducing vibrations in quartz. Any healing alleged to result from wearing crystals can be attributed to the same factors that apply to other forms of pseudoscientific alternative medicine (placebo effects, etc.).

Homeopathy

Through trial and error, and then careful monitoring of the relation between a drug and human health, modern medicine has assembled a wide variety of pharmaceuticals for ailments ranging from headaches and infectious diseases to cancer and mental illness. Its search for new drugs is a continuing one. Many substances are tested each year, but few pass the stringent tests required before a drug can be placed on the market.

Pharmaceuticals relieve pain and cure illnesses by interacting with substances in or on the human body. For example, certain chemicals that are more toxic to disease organisms than to human cells are used to control or cure infectious diseases. A problem with many drugs is that they have undesired and sometimes unpredicted side effects. Chemicals that are more toxic to disease organisms than to human cells can be toxic enough to human cells to cause significant discomfort or even do actual harm. This problem would be alleviated if people were able to receive the health benefits conferred by traditional pharmaceuticals without ingesting more than a minute dose of a substance.

Homeopathy is based on such a claim, the pseudoscientific idea that extremely tiny doses of substances that cause disease symptoms in a healthy individual can cure people suffering from similar symptoms. Furthermore, homeopathic medicine believes that the smaller the dose, the more powerful its medicine. Starting with a substance that causes symptoms of a certain disease in a healthy person, the practitioner dilutes (and then vigorously shakes) it to such an extent that, in many homeopathic medicines, *not even a single molecule of the substance remains.*

Practitioners contend that it doesn't matter if not a single molecule of the active substance remains. They speculate that the vigorous shaking of the water and alcohol mixture "charges" the entire volume of the liquid in which dilution took place to somehow "remember" that the substance was once there. There is no evidence for such a memory!

Homeopaths do negligible direct harm since the doses they give are unlikely to have any effect on anyone. The real danger comes from not getting treatments that *do* improve a person's health. Supposed cures can be attributed to the same factors that apply to other forms of pseudoscientific alternative medicine.

Hoaxes and Hoaxers

Sometimes the road to illusion is created by hoaxers, people who engage in deliberate acts of trickery with the aim of proving how gullible other people can be when a skillful imposture is presented. Following are accounts of three infamous hoaxes. The first took place about 130 years ago, the second in the early twentieth century, and the third just a few years ago.

Cardiff Giant

Whereas creatures such as Bigfoot and the Loch Ness Monster may or may not be hoaxes, the Cardiff Giant certainly was a hoax. In 1869, a statue alleged to be a 10-foot petrified prehistoric man was "discovered" by well diggers who were led to an area on a farm near Cardiff, New York. The genesis of the naked and anguished-looking stone giant was not prehistoric, but rather the idea of George Hull. One year earlier, Hull had a block of gypsum shipped from Fort Dodge, Iowa, to Chicago, carved in the shape of a human figure, and then buried on the farm. After being unearthed, the statue was placed on public exhibition in a large tent. For a fee, people were permitted to view the giant, hear a 15-minute talk, and have questions answered. Exhibition of the statue in this and other locations earned its owners a sizable amount of money.

Business boomed until an expert in the study of fossils and ancient life-forms, Othniel C. Marsh, examined it. He convinced others of the hoax by pointing out fresh tool marks and the presence of smooth, polished surfaces, which would have been roughened by any lengthy burial. Shortly thereafter, some quarrymen in Iowa recalled selling a large block of Iowa gypsum to Hull about two years earlier. It was also recalled that a very large wooden box had been hauled over backroads south of Cardiff by wagon the previous year. Later, two men from Chicago claimed to be the people who carved the giant.

Why did Hull perpetrate this hoax? For fun and for profit! After an argument with a clergyman over the phrase in Genesis, "there were giants in the earth in those days," he decided to create, bury, and then discover one of these "giants" in the earth.

How did the public react to exposure of the hoax? Exposure did little to dampen the public's fascination with the Cardiff Giant. It was moved to New York City, where it drew large audiences. After his offer to lease the statue was refused, P. T. Barnum had his own copy carved and put on display. Barnum's imitation drew even bigger crowds than the fake—a tribute to Barnum's promotional skills. After being taken on tour, and then exhibited in several other locations, the Cardiff Giant ended up in an outdoor exhibition area at the Farmer's Museum in Cooperstown, New York, where it continues to attract attention.

Piltdown Hoax

The Piltdown Man is one of the longest held and most deceptive hoaxes in paleontology (the study of fossils and ancient life-forms). This bogus species of prehistoric man, whose fossil remains were found on Piltdown Common near Lewes, Sussex, England, in 1912, was not proved fraudulent until 1953. The forgery was carefully tailored to withstand scientific scrutiny. It appeared to belong to a single creature who had a human cranium and an orangutan jaw. Its existence raised questions about the accepted ancestry of modern humans. To those who accepted its authenticity, it was an anomaly in the fossil record.

Piltdown Man was finally exposed as a forgery after a thorough reinvestigation of the fossil by anthropologists Joseph Weiner and Kenneth Oakley. Weiner and Oakley showed that the remains were a fusion of a fairly recent human cranium and an orangutan jaw with filed-down teeth to simulate the human mode of flat wear. The assembly had been stained with chemicals to simulate a prehistoric origin.

After the hoax was exposed, and the seeming anomaly resolved, the search was on for its perpetrator. For many years, the chief candidate was Charles Dawson, an amateur archaeologist who brought in the first cranial fragments from Piltdown. In the 1990s, however, the spotlight shifted to zoologist Martin Hinton when a trunk with his initials was found in the attic of London's Natural History Museum. The trunk contained bones stained and carved in the same way as the Piltdown fossils.

Crop Circles

What is usually circular, sometimes intricate and complex, created by swirling and flattening crops such as wheat without breaking the stems, frequently 36 or more feet across, and almost always made at night and discovered in the daylight? These are the crop circles that have appeared with increasing frequency since 1980, when one was found in the British fields of Wiltshire. More than 300 were found in 1991. Crop circles have been found in the United States, Canada, Japan, Germany, and Australia. In previous centuries, this effect would probably have been attributed to the devil. Nowadays, aliens are receiving credit.

It has been demonstrated that a crop circle can be created in less than an hour by a team of men using only a rope and stake. After inserting the stake into the ground, they stretch out and drag the rope over green crops (that do not break in the stem region).

Most of the crop circles are found in England, and it is there, in 1991, that two men in their sixties, David Chorley and Doug Brower, admitted to being the hoaxers responsible for many of the British crop circles. They demonstrated their skills by fooling at least one expert investigator.

Holocaust Denial: History or Pseudohistory?

The road to illusion is a slippery and dangerous slope. Failure to challenge the con artists, charlatans, and demagogues who encourage people to travel it has had disastrous consequences. For example, the racial ideas that fueled Adolf Hitler's evil Nazi empire were fashioned in the laboratory of pseudoscience. Hitler was well aware that if a misconception or a lie is repeated often enough, people will accept it as fact.

History tells the story of events such as those for which Hitler and his evil empire were responsible. History is above all a narrative. It tells the story of events that have happened. It is also a science in the sense that it asks questions about those events, and seeks to explain their underlying causes using empirical data. Like all scientists, historians take into account whatever facts are available, and then formulate theories that seem to fit the facts. If subsequent information contradicts a theory, the theory is revised or rejected. Pseudohistorians, on the other hand, decide what they want the "facts" to be, in order to support a theory they *prefer* to be true.

Historians have eyewitness testimony and documentary and physical evidence of barbaric Nazi persecution and attempted extermination of Jews and other minorities during the 12 years from 1933 to 1945. They refer to this period as the Holocaust (Shoah, in Hebrew). Because the events that took place during the Holocaust are so horrible and incomprehensible, historians are still struggling to explain them.

Pseudohistorians seeking to deny that the Holocaust ever took place (or claim that its horrors have been exaggerated by allegedly biased historians) begin with their own theories, and then try to show why the eyewitness testimony and documentary and physical evidence cited by historians must be false. For example, these people claim that the Nazi regime could not have used gas chambers to carry out a massive extermination program against Jews and others. To support this claim, they say that there was not enough residue of lethal cyanide gas in the Auschwitz gas chambers to be consistent with the amount needed for mass gassing. Among the experiments cited by the deniers to support this argument is one in which the amount of cyanide needed to kill lice was determined. The deniers take the results of this experiment and then go on to assume that it would have taken just as much cyanide to kill people as it took to kill lice. Scientists, however, have shown that it would take much less cyanide to kill humans than lice, and that it would take a much longer exposure to kill lice than to kill humans.

Not only do Holocaust deniers seek to establish "facts" that are consistent with their theory, they also discount facts that are inconsistent with their theory. Documents cited by historians are denounced as forgeries or said to mean something other than what they clearly seem to mean. Eyewitnesses to events that contradict their theory are accused of lying, being mistaken, crazy, or victims of coercion.

The reality of the Holocaust is commemorated on Holocaust Day, which is observed in Israel on Nisan 27 and elsewhere on April 19 or 20. The date is considered to be the anniversary of the thoroughly documented Warsaw Ghetto Uprising, an event that took place during April and May of 1943. During this period, the last 60,000 of the 400,000 Jews kept in this ghetto resisted the German deportation order and held out for nearly a month against the heavily armed German troops who planned to transport them to the gas chambers that Holocaust deniers refuse to acknowledge.

At the Crossroads

In marked contrast to the road to illusion, the road to reality has as its destination an accurate view of reality instead of a deluded one. It is therefore more likely to lead to successful results. Hopefully, this open road, on which ideas are freely exchanged, will become a golden road leading to relief from the backwardness and devastating poverty that plague our world.

Epilogue

Responsible believing is a skill that can be maintained only through constant practice. And since responsible believing is a prerequisite for responsible acting, we have a duty to foster that skill.

W. K. Clifford

We wish we had had the opportunity to recommend the following activities to the 39 members of the Heaven's Gate cult who chose to commit mass suicide.

- Make sure intellectual endeavors that claim to be scientific are not filled with the flaws that characterize pseudoscience.
- Read books and articles by writers such as Carl Sagan, James Randi, Stephen J. Gould, Marcia Bartusiak, Lewis Thomas, Robert Hazen, K.C. Cole, May Berenbaum, Isaac Asimov, Lynn Margulis, and James Trefil.
- Read the science sections of periodicals such as *Time* and *Newsweek* or newspapers such as the *New York Times.*
- Subscribe to a science magazine such as *Scientific American, Discover, The Sciences.*

- Read a science magazine at your local library or on the World Wide Web, for example, *Science News* at www.sciencenews.org.
- Surf the World Wide Web to visit science Web sites such as www.targetmarketing.org/course/sci/sci.htm, which provides information about science resources for educators at all levels and includes aeronautics, agricultural science, anthropology, archaeology, biology, chemistry, environmental science, genetic engineering, geology, meteorology, oceanography.
- Visit antipseudoscience Web sites such as www.csicop.org, the Committee for the Scientific Investigation of Claims of the Paranormal; www.quackwatch.com, which probes claims of alternative health practitioners; www.randi.org, the James Randi Educational Foundation, which debunks psychic feats such as reading minds and bending spoons.
- Watch TV shows such as *The World of National Geographic, Nature, Scientific American Frontiers,* or *NOVA.*
- Visit a hall of science, museum of science, zoo, aquarium, exploratorium, planetarium. For information about hands-on science museums and centers around the world, visit www.astc.org. Click "Science Center Travel Guide"; then click "Quick List of Member Web Sites."
- Accompany a naturalist on a nature walk or a geologist on a field trip.
- Enroll in a science telecourse or on-campus course at your local college.

The questioning mind is one of the most valuable assets one can have in life. It would be wise, therefore, to heed Aristotle's advice: *"If a man wishes to educate himself he must first doubt, for in doubting, he will find the truth."*

"...NO, HE CAN'T REALLY FLY... NO, THE BAD GUYS REALLY DON'T HAVE A RAY GUN... NO, THIS CEREAL REALLY ISN'T THE BEST FOOD IN THE WHOLE WORLD... NO, IT WON'T REALLY MAKE YOU AS STRONG AS A GIANT..."

"I THINK YOU SHOULD BE MORE EXPLICIT HERE IN STEP TWO."

Glossary

Abominable Snowman—unsubstantiated creature; also known as Bigfoot, Yeti, Meh-Teh, and Sasquatch

abracadabra—a magical word often appearing on an amulet

acupuncture—traditional Chinese medical technique involving insertion of fine needles into "acupuncture points"

Adamski, George—author of books about his experiences traveling into outer space with extraterrestrials

adept—a person said to be skilled at using magical or occult powers

aeromancy—divination from cloud shapes, comets, spectral formations, or other phenomena not normally visible in the sky

Agpaoa, Tony—Filipino practitioner of psychic surgery

alchemy—medieval art, the goals of which were to transmute base metals into gold, prolong life indefinitely, and manufacture artificial life

alectryomancy—divination in which a bird, usually a black hen or a white gamecock, is allowed to pick grains of corn from a circle of letters, thus forming words or names with prophetic significance

aleuromancy—divination involving slips of paper rolled in balls of dough, baked, and then mixed; one is chosen at random and presumably will be fulfilled; modern "fortune cookies" are a survival of this ancient ritual

alomancy—divination by salt

Alpha Project—study in which Michael Edwards and Steve Shaw used conjuring techniques to convince parapsychologists that they had ESP and PK abilities

alphitomancy—divination using special cakes that are digestible by persons with a clear conscience, but distasteful to others

ancient astronauts—visitors from other star systems alleged to have helped in the development of early civilizations

angel—an immortal, spiritual being

anthropomancy—divination by tearing open living human beings and examining their entrails

apantomancy—forecasts from chance meetings with animals, birds, and other creatures

apport—a substance or object brought by apparently supernatural forces into a seance room

archangel—a celestial being next in rank above an angel

Arigo, Jose—Brazilian practitioner of psychic surgery

arithmancy—an ancient form of divination involving numbers and letter values

Armageddon—the scene of a final battle between the forces of good and evil, prophesied in the Bible to occur at the end of the world

ascended master—an adept who teaches from another astral plane of existence

astraglomancy—divination involving crude dice-bearing letters or numbers

astral body—a duplicate of the human body that leaves and then returns to the body

astral plane—a dimension that exists in parallel with the real world

astral projection—out-of-body travel via astral planes

astrology—the study of the positions and aspects of heavenly bodies with the aim of predicting their influence on the course of human affairs

Atlantis—a legendary island in the Atlantic west of Gibraltar, said by Plato to have sunk beneath the sea

augury—the general art of divination

aura—“field” surrounding humans that is only visible to gifted psychics

austromancy—divination by the study of the winds

automatic writing—alleged phenomenon in which messages from other persons or entities are transmitted through an operator who writes them down “automatically”

axiomancy—divination requiring an ax or hatchet, which answers questions by its quivers when driven into a post

banshee—a female spirit in Irish folklore believed to presage a death in the family by wailing outside the house

Beelzebub—the Devil

belomancy—divination involving the tossing or balancing of arrows

Bermuda Triangle—the triangular area in the Atlantic Ocean whose corners are located approximately at Puerto Rico, the Bahamas, and the tip of Florida said to spell doom for those who venture into its domain

bezoar—a reddish stone found in the entrails of animals and used as an amulet or charm

bibliomancy—divination by books

Bigfoot—unsubstantiated creature, also known as the Abominable Snowman, Yeti, Meh-Teh, and Sasquatch

bilocation—power by which a person or object can exist in two places at the same time

biorhythms—a pseudoscientific combination of three cycles said to start at "zero" at the moment of birth, and continue like clockwork throughout life

Black Art principle—production of the illusion of floating objects by covering supports and personnel in black material and operating against a black backdrop

black magic—a form of magic performed for evil purposes

Blavatsky, Helena Petrovna—Ukrainian psychic said to be able to move objects by psychokinesis; founder of the Theosophical Society

blindfold vision—conjuring trick in which a blindfolded performer accomplishes various feats apparently without the use of sight

Blue Book—a privately published compilation of inside information about people known to frequent seances

Borley Rectory—known as "the most haunted house in England"

bumpology—nickname for phrenology

bunyip—a monster alleged to jump out of water holes in Australia and terrify passers-by

Bux, Kuda—a Kashmiri mentalist famous for his blindfold act

capnomancy—the study of smoke rising from a fire

cartomancy—fortune-telling with cards

catoptromancy—an early form of crystal gazing, utilizing a mirror that was turned to the Moon to catch the lunar rays

causimomancy—divination involving objects placed in a fire; if they fail to ignite, or burn more slowly than anticipated, it becomes a good omen

Cayce, Edgar—Cayce is called "the sleeping prophet" because when individuals would come to him with a question, he would close his eyes and appear to go into a trance; while in that state he responded to virtually any question.

cephalomancy—divinatory procedures with the skull or head of a donkey or goat

chakra—one of seven "points of power" located in the human body, to and from which psychic forces flow

channeling—transmitting information from a deceased person said to inhabit the channeler's body

charm—anything that is worn for its supposely magic effect; an amulet

chirognomy—study of traits through general hand formation

chiromancy—the study of the mystical significance of the shape and lines, markings, and developed areas of the hand; palmistry

clairaudience—claimed psychic power by which information from occult sources is "heard"

clairvoyance—claimed psychic power by which information from occult sources is "seen"

cleromancy—a form of lot casting, akin to divination with dice but simply using pebbles or other odd objects, often of different colors, instead of marked cubes

Clever Hans—a horse that was apparently able to perform mathematical calculations but was really being cued by its owner

clidomancy—divination using a dangling key that answers questions

cold reading—using cues from a person being "read" to obtain information about that person

compass trick—use of a concealed magnet to cause a magnetic compass to deflect from its normal north-south orientation

Conan Doyle, Sir Arthur—author of the Sherlock Holmes mysteries and firm believer in spirit mediums

confidence man—a dishonest person who cheats victims after first gaining their confidence

conjuring—the art of seeming to perform magic

Cottingly fairies—photographed cutouts of fairies alleged to be "real" fairies by two girls in Cottingly Glen

coven—a group of witches

critomancy—the study of barley cakes, in hope of drawing omens from them

cromniomancy—divination using onion sprouts

crop circles—patterns in grain crop fields that are created by humans but alleged to be the creation of space aliens

cryptomnesia—a phenomenon in which a subject recalls a seemingly forgotten memory and incorporates it into the present

crystal-ball gazing—divination by gazing into a crystal usually made of quartz

cyclomancy—divination from a turning wheel

dactylomancy—divination involving a dangling ring indicating words and numbers by its swings

daphnomancy—divination that requires listening to laurel branches crackling in an open fire; the louder the crackle, the better the omen

demon—a devil or evil being
demonology—the study of demons
demonomancy—divination through the aid of demons
dendromancy—divination associated with the oak and mistletoe
Devil—the major spirit of evil, ruler of Hell, and foe of God, often depicted as a man with horns, a tail, and cloven hoofs; Satan
Devil's mark—a mark placed by Satan on witches
divination—the art of foretelling future events by augury or alleged supernatural agency
Dixon, Jeane—famous seer who also claims healing power
doppelganger—a ghostly double of a living person
dowsing—using a divining rod to find underground water or minerals; "water witching"
Dunninger, Joseph—famous American mentalist
ectoplasm—a substance materialized by a spirit medium during a seance
E rays—alleged radiation emitted from unknown sources deep in the ground and said to cause cancer
evil eye—a glance by certain individuals said to induce curses and even death
exorcism—the act of expelling an evil spirit
extrasensory perception—psychic abilities to perceive that are not attributable to the normal senses
fairy—a tiny supernatural being in human form
fakir—a person in India, Pakistan, and Sri Lanka who performs conjuring tricks to earn a living
familiar—a demon in the form of an animal that acts as a companion and assistant to a witch or magician
feng shui—an ancient Chinese philosophy for creating harmonious environments

fire walking—walking barefoot over hot embers and emerging unharmed

flying saucer—a term used to describe an unidentified flying object

fortune-telling—using one of a variety of divining techniques to describe coming events

Fox sisters—three sisters who fabricated supernatural experiences and reported them as real

ganzfeld experiment—an experiment in which sensory deprivation is used to reduce possible impediments to receiving psychic information

garlic—herb said to provide protection from witches, demons, and vampires, as well as the evil eye

Geller, Uri—Israeli "psychic superstar"

geloscopy—the art of divination from the tone of someone's laughter

genethlialogy—calculation of the future from the influence of the stars at birth

geomancy—interpretation of random dots made with a pencil

ghost—the spirit of a dead person, supposed to haunt living persons or former habitats

ghoul—an evil spirit supposed to plunder graves and feed on corpses

golem—an artificially created human being endowed with life by supernatural means

graphology—the analysis of character through handwriting

gyromancy—divination performed by persons walking in a circle marked with letters, until they became dizzy and stumbled at different points, thus "spelling out" a prophesy

hand of glory—a pickled and dried hand, cut from one who has been hanged, and used in casting spells and finding buried treasure

hippomancy—a form of divination from the stamping and neighing of horses

homeopathy—healing technique in which extremely tiny doses of substances that produce disease symptoms in a healthy person are claimed to cure people suffering with similar symptoms

horoscope—an astrological chart of the zodiac

horoscopy—the casting of an astrological horoscope

hot reading—using specific, hard information obtained about a person to "read" that person

Houdini, Harry—famous escape artist and magician who devoted the latter part of his life to exposing conjurers who claimed to have supernatural abilities

hydromancy—divination by water by examining its color, ebb and flow, or the number of ripples produced by pebbles dropped into a pool, an odd number being good, an even number, being bad

I Ching—divination by reading patterns formed by thrown sticks or coins

ichthyomancy—divination involving fish

ideomotor effect—unconscious movements of a person's hands that cause movements attributed to supernatural forces

immortality—the state of eternal life

imp—a small or juvenile demon

incantation—ritual recitation of verbal charms or spells to produce a magical effect

incubus—an evil spirit believed to descend upon and have sexual intercourse with sleeping women

intuition—the act or faculty of knowing without the use of rational processes

iridology—medical diagnosis by examination of patterns in the iris of the eye

kabala—the study of controlling spirits and demons based on interpretation of the Hebrew Scriptures; also cabala

karma—the sum and the consequences of a person's actions during the successive phases of life, regarded as determining the person's destiny

Kirlian photography—photography in which photographed objects are seen surrounded by a halo-like corona or "aura"

Kreskin—stage name of famous mentalist George Joseph Kresge Jr.

lampadomancy—divination using portents from lights or torches

lecanomancy—divination involving a basin of water

Lemuria—a mythical ancient continent

levitation—an act in which the human body rises and remains above the ground

libanomancy—divination requiring incense as a means of interpreting omens

lithomancy—divination utilizing precious stones of various color; these are scattered on a flat surface, and whichever relects the light most vividly fulfills the omen

Loch Ness Monster—mythical creature alleged to inhabit a lake in Scotland; sometimes referred to as "Nessie"

lycanthropy—the condition of being a werewolf

mandala—in Oriental art and religion, any of various designs symbolic of the Universe

mantra—a sacred word or words repeated in prayer and incantation

map dowsing—using a divining rod over a map to locate items in regions of the map

margaritomancy—a procedure utilizing pearls that are supposed to bounce upward beneath an inverted pot if a guilty person approaches

materialization—appearance during a seance of an apport or ectoplasm

medium—a person thought to have powers of communicating with the spirits of the dead

mentalist—a performer who employs conjuring tricks to create effects that he claims result from psychic forces

metagnomy—divination view during a hypnotic trance

meteoromancy—divination involving omens dependent on meteors and similar phenomena

metoposcopy—reading of character from the lines of the forehead

molybdomancy—divination drawing mystical inferences from the varied hissings of molten lead

Mu—a mythical "lost" continent

Murphy, Bridey—a person alleged to have died in 1864 and been reincarnated as Virginia Tighe

myomancy—prophesy involving rats and mice, the cries they give and the destruction they cause

N rays—alleged rays discovered by French scientist Rene Blondlot and later shown to have been a product of his imagination

necromancy—divination using information obtained from the dead

New Age—a term used to denote currently fashionable ideas propounded by mystics, psychics, and gurus

Nostradamus—sixteenth century physician who wrote verses that others have interpreted as prophesies

numerology—forecasts derived from information obtained in a person's name and date of birth

occulomancy—a form of divination from the eyes

occult—of, pertaining to, dealing with, or knowledgeable in supernatural influences

oinomancy—utilization of wine to determine omens

omen—a prophetic sign

oneiromancy—use of interpretation of dreams as divination

onychomancy—the study of fingernails in the sunlight, looking for any significant symbols that can be traced

oomantia—ancient method of divination by eggs

ornithomancy—divination concerned with omens gained by watching the flight of different birds

Ouija board—a board that has all 26 letters of the alphabet and the numbers from 0 to 9 drawn on it; questions are asked of the board, and the answers are provided by spirits that guide the hands of people to specific letters or numbers.

out-of-body experience—an event in which a person in some way leaves her body and then perceives various sights and sounds.

palmistry—the study of the mystical significance of the shape and lines, markings, and developed areas of the hand; chiromancy

paranormal—not within the range of normal experience or scientifically explainable phenomena

parapsychology—the study of reported but unsubstantiated events that have no presently known explanation

pegomancy—divination requiring spring water or bubbling fountains

pendulum—a divination device consisting of a mass suspended at the end of a thread so that the mass is free to swing in a vertical plane

philosopher's stone—device by which base metals can be transmuted into gold

phrenology—determination of a person's character traits by studying bumps on the person's head

phyllorhodomancy—an intriguing form of divination dating from ancient Greece; it consists of slapping rose petals against the hand and judging the success of a venture according to the loudness of the sound

placebo effect—the ability of the body to heal itself as a result of belief in a cure

planchette—a heart-shaped device used to point to letters and numbers on the Oiuja board

police psychics—people that claim to be able to use psychic abilities to help police solve crimes

poltergeist—a noisy ghost

possession—occupation of a person's body by a devil, demon, or spirit

potion—an ingested substance made to serve a magical function

precognition—knowledge of a future event or circumstance that is obtained by paranormal means

premonition—a foreboding of the future

prophesy—the general ability to foretell future events

psi—a word used to denote paranormal phenomena

psi gap—the supposed discrepancy between the United States and the Soviet Union in the use of psychic phenomena as a defense mode

psychic—a person said to possess extraordinary, especially extrasensory and nonphysical, mental processes

psychic portraits—"portraits" of dead persons that are produced through the use of psychic powers

psychic surgery—a pseudoscientific procedure in which the practitioner claims to cure organic ailments by parting and reaching through the skin with his bare hands to remove tissue often claimed to be cancerous

psychokinesis—the production of motion, especially in inanimate and remote objects, by the exercise of psychic powers

psychometry—the detection of "psychic vibrations" that have been absorbed by objects and by places

pyramid power—"energies unknown to science" associated with pyramidal shapes

pyromancy—divination by fire

qi—a "life force" believed to circulate in the body through pathways called meridians

Rampa, T. Lobsang—pen name for an Englishman who claimed to be gifted with mystic powers because his body was occupied by a Tibetan whose "third eye" was opened when a hole was poked in the Englishman's forehead

rapping—tapping signals originating with spirits

reincarnation—being reborn in another body

remote viewing—a phenomenon in which a psychic is able to obtain information about a distant location

rhabdomancy—divination by means of a wand or stick

rhapsodomancy—divination performed by opening a book of poetry and reading a passage at chance, hoping it will prove to be an omen

runes—specially inscribed dice that are thrown for divination purposes

Sai Baba—an Indian yogi who uses conjuring techniques to convince followers that he has miraculous powers

Sasquatch—unsubstantiated creature, also known as the Abominable Snowman, Bigfoot, Yeti, and Meh-Teh

sciomancy—divination gained through spirit aid

scrying—divination of past or future events by gazing into a crystal, mirror, or bowl of water

seance—a meeting of persons to receive spiritualistic messages

second sight—a phenomenon in which two people are seemingly able to know each other's thoughts

Shroud of Turin—a woven cloth purported to be the burial cloth or shroud of Jesus Christ

sideromancy—the burning of straws on a hot iron and studying the figures thus formed, along with the flames and smoke

sitter—a person who participates at a seance

sorcery—the use of magic methods to obtain power over others

sortilege—the casting of lots in hope of a good omen

spell—an incantational word or formula

spirit—any supernatural being, such as a ghost

spirit guide—a spirit that a spirit medium says is serving as a go-between with the "other world"

spirit medium—a person who claims to be able to call up spirits

spirit photography—photography that supposedly captures the image of the spirit of a person who has died

spodomancy—divination provided by omens from cinders or soot

spontaneous human combustion—the supposed process in which a human body suddenly bursts into flames as a result of heat generated by internal chemical action

sprite—a small or elusive supernatural being; an elf or pixie

stichomancy—divination involving the opening of a book, hoping that a random passage will give inspiration

stigmata—spontaneous wounds corresponding to the traditional wounds on the body of Jesus Christ

stolisomancy—the drawing of omens from oddities in the way people dress

sycomancy—divination performed by writing messages on tree leaves; the slower they dry, the better the omen

table tipping—an alleged phenomenon in which people place their hands on a table and "will" it to rise, tilt, or rotate

tarot cards—a special deck of 78 cards used for divination

tea-leaf reading—divination using patterns formed by tea leaves in a cup; tasseography

telepathy—the ability to perceive the thoughts or emotions of others

teleportation—the ability to transport oneself from place to place magically

tephramancy—the seeking of messages in ashes, usually of tree bark

tiromancy—an odd form of divination involving cheese

unidentified flying object (UFO)—object observed in the sky that is identified as an alien spacecraft

unicorn—mythical animal usually represented as a horse with a single spiraled horn projecting from its forehead and often with a goat's beard and a lion's tail

vampire—a reanimated corpse that rises from the grave at night to suck the blood of sleeping persons

warlock—a male witch

werewolf—a person transformed into a wolf or capable of assuming the form of a wolf at will

witch—a woman who practices sorcery or is believed to have dealings with the devil

xylomancy—divination from pieces of wood; some diviners pick them up at random, interpreting them according to their shape or formation; others put pieces of wood upon a fire and note the order in which they burn, forming conclusions as to omens, good or bad

Yeti—unsubstantiated creature, also known as the Abominable Snowman, Bigfoot, Meh-Teh, and Sasquatch

zodiac—a band of the celestial sphere that represents the path of the principal planets, the Moon, and the Sun

Additional Reading

Science

Atkins, P.W. *Creation Revisited: The Origin of Space, Time, and the Universe.* London: Penguin Books Ltd., 1994.

Brody, D.E. and Brody, A.R. *The Science Class You Wish You Had . . .: The Seven Greatest Scientific Discoveries in History and the People Who Made Them.* New York: Perigee Books, 1997.

Dennett, D.C. *Darwin's Dangerous Idea: Evolution and the Meaning of Life.* New York: Touchstone, 1996.

Derry, G.N. *What Science Is and How It Works.* Princeton, NJ: Princeton University Press, 1999.

Gribbin, J. *Almost Everyone's Guide to Science.* New Haven: Yale University Press, 1999.

Grinnelle, F. *The Scientific Attitude.* Boulder: Westview Press, 1987.

Hatton, J. and Plouffe, P.B. *Science and Its Ways of Knowing*. Upper Saddle River, NJ: Prentice Hall, 1997.

Hazen, R.M. and Trefil, J. *Science Matters: Achieving Scientific Literacy*. New York: Doubleday, 1992.

Lee, J.A. *The Scientific Endeavor: A Primer of Scientific Principles and Practice*. San Francisco: Addison Wesley Longman, 1999.

Marshall, I. and Zohar, D. *Who's Afraid of Schrodinger's Cat? All the New Science Ideas You Need to Keep Up with the New Thinking*. New York: William Morrow, 1997.

Moore, J.A. *Science as a Way of Knowing: The Foundations of Modern Biology*. Cambridge, MA: Harvard University Press, 1993.

Speyer, E. *Six Roads from Newton: Great Discoveries in Physics*. New York: John Wiley & Sons, 1994.

Spielberg, N. and Anderson, B.D. *Seven Ideas That Shook the Universe*, 2nd ed. New York: John Wiley & Sons, 1995.

Stanovich, K.E. *How to Think Straight About Psychology*, 5th ed. New York: Longman, 1998.

Wynn, C.M., Wiggins, A.W., and Harris, S. *The Five Biggest Ideas in Science*. New York: John Wiley & Sons, 1997.

Science Versus Pseudoscience

Aaseng, N. *Science Versus Pseudoscience*. New York: Franklin Watts, 1994.

Della Sala, S., ed. *Mind Myths: Exploring Popular Assumptions About the Mind and Brain*. New York: John Wiley & Sons, 2000.

Friedlander, M.W. *At the Fringes of Science*. Boulder: Westview Press, 1995.

Gardner, M. *The New Age: Notes of a Fringe Watcher*. Buffalo, NY: Prometheus Books, 1988.

Gardner, M. *Weird Water and Fuzzy Logic: More Notes of a Fringe Watcher.* Amherst, NY: Prometheus Books, 1996.

Hess, D. *Science in the New Age: The Paranormal, Its Defenders and Debunkers, and American Culture.* Madison: University of Wisconsin Press, 1993.

Hines, T. *Pseudoscience and the Paranormal: A Critical Examination of the Evidence.* Buffalo, NY: Prometheus Books, 1988.

Krauss, L. *Beyond Star Trek.* New York: Basic Books, 1997.

Park. R.L. *Voodoo Science: The Road from Foolishness to Fraud.* New York: Oxford University Press, 2000.

Randi, J. *An Encyclopedia of Claims, Frauds, and Hoaxes of the Occult and Supernatural.* New York: St. Martin's Griffin, 1997.

Randi, J. *Flim-Flam? Psychics, ESP, Unicorns, and Other Delusions.* Buffalo, NY: Prometheus Books, 1982.

Sagan, C. *The Demon-Haunted World: Science as a Candle in the Dark.* New York: Random House, 1996.

Schick, T., Jr., and Vaughn, L. *How to Think About Weird Things: Critical Thinking for a New Age.* Mountain View, CA: Mayfield, 1995.

Shermer, M. *Why People Believe Weird Things: Pseudoscience, Superstition, and Other Confusions of Our Time.* New York: W.H. Freeman, 1997.

Stein, G. *Encyclopedia of Hoaxes.* Detroit: Gale Research, 1993.

White, M. *Weird Science.* New Haven: Yale University Press, 1999.

UFOs and Alien Abductions

Achenbach, J. *Captured by Aliens: The Search for Life and Truth in a Very Large Universe.* New York: Simon & Schuster, 1999.

Davies, P. *Are We Alone?* London: Penguin, 1995.

Dick, S. *Life on Other Worlds: The 20th Century Extraterrestrial Life Debate.* New York: Cambridge University Press, 1998.

Frazier, K., ed. *The UFO Invasion: The Roswell Incident, Alien Abductions, and Government Coverups.* Buffalo, NY: Prometheus Books, 1997.

Klass, P.J. *Bringing UFOs Down to Earth.* Buffalo, NY: Prometheus Books, 1997.

Randle, K.D., Estes, R., and Cone, W.P. *The Abduction Enigma: The Truth Behind the Mass Alien Abductions of the Late 20th Century.* New York: Tom Doherty Associates, 1999.

Randles, J. *Men in Black: Government Agents or Visitors from Beyond—Finally the Truth.* New York: St. Martin's Mass Market Paper, 1997.

Randles, J. and Hough, P. *The Complete Book of UFOs: An Investigation into Alien Contacts and Encounters.* New York: Sterling, 1996.

Out-of-Body Experiences and Entities

Blackmore, S.J. *An Investigation of the Out-of-the-Body Experience.* London: Heinemann, 1982.

Cohen, D. *Encyclopedia of Ghosts.* New York: Dodd, Mead, 1984.

Crapanzano, V. and Garrison, V. *Case Studies in Spirit Possession.* New York: John Wiley & Sons, 1977.

Finucane, R.C. *Ghosts: Appearances of the Dead and Cultural Transformations.* Buffalo, NY: Prometheus Books, 1996.

Gordon, H. *Channeling into the New Age: The "Teachings" of Shirley MacLaine.* Buffalo, NY: Prometheus Books, 1988.

Houdini, H. *A Magician Among the Spirits.* New York: Arno Press, 1972.

Irwin, H. *Flight of Mind: A Psychological Study of the Out-of-Body Experience.* Metuchen, NJ: Scarecrow Press, 1985.

Rogo, D.S. *The Poltergeist Experience.* New York: Penguin, 1979.

Underwood, P. *The Ghost Hunter's Guide.* New York: Blandford Press, 1986.

Astrology

Bok, B.J. and Jerome, L.E. *Objections to Astrology.* Buffalo, NY: Prometheus Books, 1976.

Culver, R.B. and Ianna, P.A. *Astrology: True or False?* Buffalo, NY: Prometheus Books, 1988.

Gauquelin, M. *The Scientific Basis of Astrology: Myth or Reality?* New York: Stein and Day, 1969.

Martens, R. and Trachet, T. *Making Sense of Astrology.* Buffalo, NY: Prometheus Books, 1998.

Roszak, T. *Why Astrology Endures.* San Francisco: Robert Briggs Associates, 1980.

Stewart, J.V. *Astrology: What's Really in the Stars.* Buffalo, NY: Prometheus Books, 1996.

Evolution and Creationism

Asimov, I. *In the Beginning . . . Science Faces God in the Book of Genesis.* New York: Crown, 1981.

Berra, T.M. *Evolution and the Myth of Creationism: A Basic Guide to the Facts in the Evolution Debate.* Stanford, CA: Stanford University Press, 1990.

Eldredge, N. *The Triumph of Evolution.* New York: W.H. Freeman, 2000.

Eve, R.A. and Harrold, F.B. *The Creationist Movement in Modern America.* Boston: Twayne, 1991.

Hanson, R.W., ed. *Science and Creation: Geological, Theological, and Educational Perspectives.* New York: Macmillan, 1986.

Kitcher, P. *Abusing Science: The Case Against Creationism.* Cambridge, MA: MIT Press, 1982.

Pennock, R.T. *Tower of Babel: The Evidence Against the New Creationism.* Cambridge, MA: MIT Press, 1999.

Ridley, M. *Evolution.* Boston: Blackwell Scientific, 1993.

Shermer, M. *How We Believe: The Search for God in an Age of Science.* New York: W.H. Freeman, 1999.

Strahler, A.N. *Science and Earth History: The Evolution/Creation Controversy.* Buffalo, NY: Prometheus Books, 1987.

ESP and Psychokinesis

Alcock, J.E. *Science and Supernature: A Critical Appraisal of Parapsychology.* Buffalo, NY: Prometheus Books, 1990.

Braude, S. *The Limits of Influence: Psychokinesis and the Philosophy of Science.* New York: Methuen, 1986.

Gardner, M. *How Not to Test a Psychic: Ten Years of Remarkable Experiments with Renowned Clairvoyant Pavel Stepanek.* Buffalo, NY: Prometheus Books, 1989.

Gordon, H. *Extrasensory Deception.* Buffalo, NY: Prometheus Books, 1987.

Hansel, C.E.M. *The Search for Psychic Power.* Buffalo, NY: Prometheus Books, 1989.

Keene, M.L. *The Psychic Mafia.* Buffalo, NY: Prometheus Books, 1996.

Nickell, J., ed. *Psychic Sleuths: ESP and Sensational Cases.* Buffalo, NY: Prometheus Books, 1991.

Randi, J. *The Magic of Uri Geller.* New York: Ballantine, 1975.

Stenger, V.J. *Physics and Psychics: The Search for a World Beyond the Senses.* Buffalo, NY: Prometheus Books, 1990.

Index

C

D

E

F

G

H

O

P

Q

R